Peak Water

Peak Water

How the control of water gave us civilisation
and why it is now failing

ALEXANDER BELL

Luath Press Limited

EDINBURGH

www.luath.co.uk

The paper used in this book is acid-free, recyclable and biodegradable.
It is made from low chlorine pulps produced in a low energy, low
emission manner from renewable forests.

Printed and bound by
Exacta Print, Glasgow

Typeset in 11 point Sabon by 3btype.com

Contents

Foreword

WE ONLY NEED to build a sandcastle and flood its moat with seawater to know the deep pleasure of changing water's course. The fact that the tide will wash it all away does not take the edge off a summer's day. This book is about how mankind has played with water to the point that all the sandcastles of cities and civilisation may be about to go.

This isn't a science book, or dry history, but a story telling how man made the leap into civilisation by controlling water 6,000 years ago. This great discovery gave rise to the first sustainable cities. From there, the control of water became essential for the rise and fall of empires and the spread of ordered, civilised living.

In the great sweep of human history, civilisation has lasted no more than cigarette burn at the end of a long party. Water control didn't just start this brief but important phase in human history, it runs through it, like an underground river, which floods and bursts beneath all that we do. All the large and small innovations and discoveries that ensure you can read this book in the safety of your home, guaranteed a clean water supply, safe in the knowledge that there will be food in the shops tomorrow, even if your fridge is empty today, can be credited, at least in part, to water control.

This is not a book on the biological or scientific qualities of water, amazing as those are. It doesn't trouble with the details of the molecule that is two hydrogen atoms and one oxygen atom[1]. Nor do we range over the significance of salt water, as the birthplace of life, or sea trade that did so much to develop human understanding. This is about how civilisation is thirsty. It has relied on pumping up groundwater and redirecting rivers, on dams and drainage, for its success. We got fresh water to follow us, and built astonishing things.

This is a short book written in a simple style. It is designed for the general reader. The ideas here are for everyone; I would like any citizen of the planet to have a good chance of joining in this debate. It is about our world and how we have shaped it, and it is about our future.

I'm a writer, not an academic. To the many more learned people than me who have written reports, books and papers on water, I am very grateful. No doubt I will have missed some helpful studies, and that is my oversight. To the academics who willingly offered their time to contribute ideas and associations, thank you. To the kind people who read this manuscript, thanks. Any errors that remain are the fault of me alone. The modern spelling of ancient names can vary from source to source but I have tried to use the most recognised versions. This does not pretend to be a comprehensive study of what is a vast field. It does hope to be a new way of looking at the world; it offers a point of view, the beginning of a fresh argument.

This book was written for Murdo and Eilidh, the start of promise.

I

THE FIRST TASTE

WATER IS A RENEWABLE RESOURCE, yet it is running out. How do we explain that puzzle? The answer lies in the way mankind has developed over the last 6,000 years. We invented and expanded civilisation, which has given us many wonderful things and allowed the population to dramatically increase. When lots of people want to enjoy a civilised lifestyle, they drain water reserves. By repeating habits begun millennia ago, we empty the wells that have made our cities and nations great. To understand this, we need to float down the river of history and watch as man learns to tap the essence of all life.

CHAPTER ONE

Where's the Water?

IT IS 46 DEGREES in the sun. The air shimmers over the sand dunes. At first it seems like a mirage, the shapes that drift in the distance, as if they are inventions of the heat. As you get closer you become convinced that something isn't right. The vague forms look like buildings, but huge ones. Vast towers linger in the dust. Closer still and doubt passes. These are real. There is a city ahead, and it has been built on a fantastic scale. This is Dubai, a metropolis that jags out of the desert like a shaft of stone. It is the most astonishing new city in the world, and it is not an invention of heat, but of oil.

When the global financial system became infected with doubt in 2007 and 2008, only one place seemed immune. The oil rich states of the Middle East appeared able to ride out the crisis. They relied on the value of crude, and it was trading at around $100 a barrel. Of these states, the most remarkable is the United Arab Emirates. The UAE has gone from poverty to fabulous wealth in a generation, thanks to the black stuff pumped up from under the sand. The monument to the nation's new glory is Dubai.

Civilisation is partly the history of cities and great buildings. Dubai's bold straining for the sky seems to be shouting how much better it is than the peaks of Manhattan, or the freeways of Los Angeles. As those cities wanted to be better than London or Paris, as ancient Rome wanted to be better than Egypt, Dubai is merely the latest urban wonder.

As the financial crisis wears on, it turns out Dubai isn't built on oil, but confidence. The neighbouring emirate, the big player in the UAE, Abu Dhabi, has the oil. The fantasy grandeur of Dubai is one of capitalism's tricks, an illusion of wealth built on speculation and debt. Somehow this makes it more admirable – a human venture propped up by ambition and faith in our ability to survive.

When you stand in 46 degree heat, in the desert, you don't think of oil or human trickery. You think of water. Money and the inventions of man evaporate from your mind. It reverts to survival mode. Water is the priority. It is the stuff of life, without which the human body dies after three days. The UAE is a sea of hot sand. In truth, this is the last place you would think to build a city. This city should really make you think – why?

This is the story of why mankind thought it smart to build huge cities in the desert, or why civilisation got so confident it thought the broiling sun and bone-dry earth was no obstacle to urban life. It is about how civilisation became so successful at controlling fresh water it never stopped to think what would happen if the water ran out.

It is also about our resources. Oil lubricates modern life and it may be running out. This is expressed as 'peak oil', which means the point at which global supplies can only diminish. I once spent an afternoon with an oil economist for an international bank. He had been told off for talking down 'peak oil'. Peak oil is good for business, as it boosts prices. The economist explained that estimates of reserves were notoriously unreliable, the statistics used to calculate peak oil were particularly flawed, and anyway mankind was unusually gifted at finding more of the stuff.

The economist would scoff at the idea of peak water. There is plenty of it. Not only that, it is renewable and inexhaustible. No matter how we use water, it will be recycled back to the sea, up to the clouds, and fall as rain again some day. But this isn't an exercise in numbers: this is about life. Peak oil isn't just about cold statistics, much as that may horrify the economist. It's a way of focusing minds on the future, when the gas station will be empty. And peak water is not about global supplies, but the fact that civilisation depends on wells that are empty, and rivers that have run dry. I don't know when the oil will run out, but we are on the brink of 'peak water'.

Dubai has the highest water consumption per capita in the world. The heat plays a part certainly, but equally important is that civilisation is very thirsty. It sucks up water for construction, agriculture and industry. It gulps in the name of luxury and cleanliness. Civilisation is wet. For man to build such monuments, to travel the amazing path of civilisation, he had to keep inventing new ways of getting fresh water, and it is the legacy of this thirst that means we are reaching peak water.

Some of the crisis is caused by global warming, but not all. If the climate didn't change one jot within the next millennia, our civilisation would still have to adapt, because we have created a civilisation based on water usage that is not sustainable.

Civilisation is not a given. Of the three and half million years that humans have existed, we have been civilised for only 6,000, of which only the last 200 years have provided anything close to what we now consider to be normal. The idea of civilisation as our default position is fiction. We live at the peak of human ingenuity, but it is a fragile state. All that we have can go in a frighteningly short period of time[2].

When Kenneth Clarke wrote a TV series and book entitled *Civilisation*, he admitted he '...had no clear idea what it (civilisation) meant, but I thought it was preferable to barbarism'. That's the point of civilisation to some – it's where man would rather be. It is a dualistic set-up; if we are not civilised, we must be barbarians.

I must tread carefully here. Clarke went on to write; 'Writers and politicians may come out with all sorts of edifying sentiments, but they are what is known as declarations of intent'. I should be careful of my own self-serving declaration on civilisation. That said, it is important to pin down the idea.

To the ideas of the city and justice, it's not unreasonable to attach notions of cleanliness, and longevity of life, and science and medicine, and good manners and rational thought, and tolerance. Couple all this to the ideas of progress and betterment, of improvement, and you have civilisation. We know Dubai is civilised, even if it isn't quite to our taste and habit.

To some, man has propelled himself into civilisation, incapable of moderating the process.

> It is tempting to think of civilisation as an end, rather than a means, of human existence. But that would be wrong. There is no primary law that drives people towards civilisation as a mode of living. Civilisation is a very impressive demonstration of human ingenuity applied to the problems of fulfilling human requirements, but, in the story of human evolution, it is actually a very recent innovation...

says the essayist Ronald Wright.

I don't assume that 'civilisation' is better than other social systems, or whether we craved civilised living, or merely stumbled into it as a convenient way of managing life in this climate[3]. What matters is that a new way of organising ourselves occurred in Mesopotamia around 4,000 BC, and we still rely on this innovation.

Cities like Dubai have been one of the great innovations of civilisation. More of us live in urban areas now than in the countryside. We huddle together in apartment blocks, swarm towards school buildings and offices, gather food on supermarket shelves, and seek happiness in the dark corners or the sunlit parks. Now let us destroy all that.

Rip up Dubai, break it down to its core parts. Beneath the stone and tarmac are wires and pipes – a knot of services that make city life possible. If you were to rip out all the telecoms cables, you'd end up with billions of miles of copper string and fibre optic and a world that could no longer talk via machines. The phones would go down and the computers would stop scattering bytes of trivia to one another. The great technological leap of the digital age would come to an end. There would be a huge price to pay for this, in the loss of services that relied on computers, but civilisation would adapt.

Strip out the electricity cables, and the city would go dark, its hum and buzz run down to an unpleasant silence. Machines would shut off, air con would stop and the signals that keep the city functioning would go black. But civilisation would adapt.

Tear up the underground tanks which store petrol for the street level pumps and there would be chaos. The city depends on a constant stream of food being trucked in, and on the economic output of car-driving workers. Also, strangely, carbon fuel seems to occupy a special place in the heart of modern citizens – a kind of liquid nutrient that feeds a sense of identity and independence. Anyway – rip it up, and civilisation would adapt.

The gas pipes, feeding the boilers and kitchen rings, would be missed no doubt, and the resulting return to open fires would be risky in a modern city, but civilisation would cope.

What would bring the city down are the remaining pipes and tunnels. The oldest hidden infrastructure of any city is the drinking water and the sewage system. Water is the first service to be provided to the citizens, and for good reason. Without it, you have no city. And for significant tracts of the globe, you would have no farming either, if the irrigation ditches were trashed.

On this simple provision civilisation was born and grew; allowing mankind to grow from a wandering, vulnerable hunter-gatherer to a clean, healthy, city-living creature functioning in a complex society. This is the age of water.

Leonardo da Vinci seemed to have civilisation in his DNA. He was curious, creative and energetic, enthralled by the mystery of nature and a pioneer of scientific enquiry. One of the many areas he studied was water. The River Arno, which runs through Florence, fascinated him. He wondered where the water came from to keep the current flowing. His notes on this laid the foundations for some modern water studies[4].

Leonardo was worried about the biblical flood. He couldn't square the sudden appearance, and disappearance, of a vast amount of water, as described in scripture, with what he knew about the real substance. Where did it all go? How did it flow away, if there was no current? Yet the idea of a deluge fascinated him. He drew images of what it might look like, capturing the sense of chaotic power that a huge body of water has[5]. As such, Leonardo identified a vital quality of water. It can be measured, and follows predictable patterns, but it is also wild. This contradiction is at the heart of our current problem.

Imagine you have a bucket. This is all the water on the planet. You cup your hands and scoop out a tiny fraction of the water, which now sits in your reservoir of flesh. That is all the fresh water on the planet. The vast majority of blue gold is salty; 97 per cent is undrinkable. With your hands full, you have the three per cent that is fresh. You hold life and civilisation. To have an idea of how much water is available to humans, you must open your hands and let the water fall. Now shake your hands, blow on them a bit – and the dampness left, the glistening traces of water in the crease of your skin, that is all the fresh water accessible to humans.

Most of the fresh water is locked in the ice caps at the poles, or is in inaccessible aquifers underground. It is estimated that of all Earth's water, you and I have a reasonable shot of reaching less than one per cent of it.

On reading this sequence, it's like learning you won the lottery, but gradually discovering that everyone called the same numbers, and your sweepstake is worth pennies. However it may be small in the scheme of things, but it should be enough.

The water available to us is the stuff captured in rainfall. This is a part of the water cycle, the process whereby water evaporates off the sea's surface and from lakes, rises into the atmosphere to form clouds, and then falls as rain. It's estimated that 30 trillion gallons of fresh water falls every day. The vast majority returns to the sea before we've had a chance to get to it; so it's reckoned that roughly a third is available to us, of which floods and inaccessible rivers deny us even more. The punchline is that around 12,500 cubic km per year are yours and mine – the traces of water caught in the creases of your palm.

Distillers call whisky that evaporates during storage the angels' share. It may not be scientifically precise, but in essence we have access to the angels' share of water that wafts off the planet's vast reserves. And it should be enough to slake our thirst, wash our skin and grow our food. In fact, we are probably only using about half of it right now. Statistically, we are good for water.

Water consumption shot up with modernity. The need for more food to feed more people, who needed more commodities, contributed to a major increase in water use. The informed guess is that we used 110 cubic km of water in 1700 AD. Three hundred years later, at the beginning of the 21st century, we are using around 5,000 cubic km.

That sounds alarming, but the use of water in the developed world is steadily declining. Our washing machines use less, the production of commodities such as steel or cars uses less and those parts of the developed world with irrigation are learning to spread the water better. In the USA, water use per capita peaked in 1980, and had dropped by a tenth by 1995.

So, while contemporary civilisation has found a multitude of new ways of using water, and uses far in excess of what any previous civilisation has used, it has also learnt to curb this consumption. However, civilisation is, in modern parlance, an aspirational brand. Thus two billion Chinese and Indians are struggling to catch-up with western economic might and living standards. This is part of the bad news.

Current estimates suggest that the world may begin to use 100 per cent of the available run off by 2035. The UN reckons we will suffer serious water shortages from 2020 onwards. So how do we tally an apparent excess of fresh water, and current usage running at around 50 per cent, coupled to a reduction in water consumption in the developed world, with water shortages? Much as Leonardo da Vinci puzzled over the biblical flood, so the modern water crisis seems to make little sense.

There is a tedious row over semantics in the world of water use. When environmentalists shout that water is running out, hydrologists and scientists yell back that it isn't. Factually, the scientists are right. If you want to know that water shortages will never be an issue in your life (not allowing for the wild card of climate change) then move to Siberia. On the shores of Lake Baikal you will witness the largest body of fresh water in the world. The trouble with Lake Baikal, and with Siberia, is that the world doesn't live there. That's the trouble at the heart of water statistics. There is plenty of water, but not in the same place as the people. That is the world's water problem.

An important part of the problem is the method by which water reserves are assessed. Water control played its part in the creation of the nation state, as we shall see, but national boundaries are an awful way of measuring water resources, as if a line of ink on a map has any relevance to a river.

Statistically, Australia is in the lucky group of nations with plentiful supplies of water. The United Kingdom has insufficient water. Better off than the UK is the USA, with relatively sufficient supplies. Go tell the people on the banks of the Murray River in southern Australia that they are the lucky ones. A drought has made the river a trickle, which no longer reaches the sea. Indeed the very pattern of settlement in Australia by the Europeans, with a few big cities and long stretches of nothing in between, is indicative of a land where water is scarce. The statistics show the nation to be water rich because of the tropical forest in the north.

The black umbrella is an icon of London. We rightly associate the UK with rain – in many parts it rains a lot. Break the UK into its constituent nations, and Northern Ireland, Scotland and Wales don't lack for water. Split England into regions, and most of them are sodden. The statistical blip putting the UK into the insufficient category is the concentration of the population in the south east of the country. The people have gathered where the water isn't. The UK has a water problem, but most of the people in the UK do not. They are much better off than their cousins in the populous parts of Australia, despite what the numbers show.

As for the USA, providing one statistical verdict on its water supplies is meaningless. The water demands of New York, Florida, New Orleans and Las Vegas couldn't be more different. In broad terms, the west of the nation has a major water shortage, while the east is okay.

Of course, one could say the same of oil reserves. The oil in the USA is to the west, the reserves in the east having been pumped dry. The oil in the UK is in Scotland, England never having had any to speak of. Yet the economy in both is oil-based, and successful. But water isn't oil. You can pump up the black stuff and put it into tankers and ship it anywhere in the world, but that doesn't work with water.

Moving water is the Faustian pact that man always regrets. It becomes a burdensome task that ultimately ruins environments, destroys productive land, wastes fantastic amounts of resources, and leads to the destruction of nations and empires. It is the folly of past civilisations. It is the flaw in man's water control plans. The lesson of history is that it is ultimately better to let the people follow the water. That is, however, a lesson no society is prepared to contemplate, until it is too late.

At this moment, Colonel Gaddafi in Libya is pumping up ancient water reserves, trapped under the Saharan sands when the region was a wet forest. This is an engineering wonder of the world. Two massive pipes, constructed by American engineers, bring cool fresh water to the fields on the Mediterranean coast. Even when the US claimed Libya a sworn enemy, it clung on to this project. You might think this shows how water needs can rise above national squabbles. How-ever, let's not forget this was a multi-billion dollar contract.

Gaddafi wanted the people to follow the water, at least a little. He planned to pump it up and irrigate the desert. The farmers didn't want to 'make the desert bloom' as they preferred living among the trappings of civilisation on the coast, so the president moved the water to them instead.

The huge concrete tubes that bore beneath the hot sand are painted green, the Islamic colour for water, and hope, and faith. The last two play a tragic role in this drama, not just in Libya but in all great water schemes. The aquifer supplying the pipes is finite. As it is, neither of the two main schemes to date has supplied the amount of water promised. The water that is coming is being used to grow water-intensive crops like wheat and citrus fruit. And the farms are a thousand kilometres from the source. During these good times, when the water lasts, no effort is being put into conservation, or growing better-suited crops, or generally planning for the future. When the aquifer dips too low, the pipes will be dry. Then the farmers, and the president, will realise that the success they thought had been delivered for all time was only a passing thing. There will be no water to sustain the fields, no crops to sell, no money for the people, and the whole glorious folly will slowly drown in the drift of sand.

Taking water from wet areas to arid zones is like pumping more fuel onto a burning oil well – pointless and wasteful. That is the trouble with water, and that is the world's trouble. For the Canadians, the north Europeans, and the Russians, for the northern Brazilians and the war-battered lands of the Congo in Central Africa there is no problem. Here is the water. But the increase in global population isn't predicted for the Congo, or Brazil, or Canada. The rise to a population of just under nine billion people by 2050 is forecast to occur in China, India, Asia, the Middle East and Africa. Thus our problem is not simply a mismatch between people and water reserves, but between population expansion and water. What do we do with all the people in lands where the water has gone?

As it is, the great rivers of the world are already running dry. The Colorado doesn't make it to the Pacific Ocean for half the year, its delta is eroding and the seawater encroaching on what was once a gush of fresh water. The rivers that made China, the Yellow and Yangtze, also fail to reach the sea for stretches of the year. For those reading in wet countries, imagine the Thames reduced to a muddy ditch outside the Palace of Westminster, or the Rhine petering out somewhere in northern France, so that children can cross the bed without fear. That's what happens to the Nile and the Ganges. The iconic rivers of our imagination, the primal flow of mankind, of civilisation, of progress and the binding myths of our existence, are drying up.

The rivers are the visible, potent, symbols of our deluded belief in water control. With them come wetlands, flood plains, natural irrigation and the steady, if slow, replenishment of underground water reserves. In the dry riverbed, in the cracked soil, see the withering of all our various water supplies. The consequence is already there to be felt.

'All peoples, whatever their stage of social development and their social and economic conditions, have the right to have access to drinking water in quantities and of a quality equal to their basic needs'. That was the goal of the first United Nations water resources conference, held in Mar Del Plata in 1977. Wrapping water in the language of rights may have comforted the delegates, and may still seem a good thing to do, but it has only gone so far in letting people drink.

By 2050, around half of the world will live in nations which are short of water. Allowing for the regional variation in water supply within these nations, water deprivation will affect billions of us. As it is, 65 per cent of the world's water reserves are in just 10 nations. Again, the Canadians and the Brazilians, through no ingenuity but the luck of geography, are water rich while the booming populations of the Middle East and Asia are dry. One fifth of the population occupy land supplying two per cent of the globe's water.

There is another problem. For civilisation to bloom, it had to tap underground water. Wells seem a benign, ancient way of getting water. Yet the bucket on the rope, or the pump sucking deep in the earth, may yet come to be seen to be as dangerous to the environment as the oil well. When we extract oil from the ground, we expect it to run out eventually. The same happens to water.

Underground water reservoirs are called aquifers. They form when rainwater seeps through the soil and collects in rocky chambers. Because rain has been falling for billions of years, these subterranean reserves can be huge. However, if the bucket is big enough, or the pump strong enough, the rate at which man takes water out will outpace the rate at which naure replaces it.

Many areas of human population are dependent on underground water. The rivers of the Middle East, the western USA or India are incapable of providing enough water for the scale of human development that exists in these regions. The cities and farms of these places grew up because water was pumped up from wells. That water is running out.

The World Health Organisation (WHO) reckons a third of the Earth's population lack the necessary quantities of water to survive, which is said to be 100 to 200 litres a day, when all agriculture, industry and domestic uses are accounted for. But WHO says half the world already suffers from poor to no sanitation, while a quarter of us have no clean water at all, contributing to one third of all fatalities in developing nations, with 80 per cent of all disease in these parts water-borne. Our inability to truly control fresh water for all is the single largest threat to the health, happiness and longevity of many. We have failed, and are set to do a lot worse.

So the statistics may give us 12,500 cubic kilometres of water, but I believe we should halve that to around 6,000 cubic kilometres, when global population spread is taken into account. It is an estimate and is unlikely to end the debate between scientists and environmentalists, but gives the rest of us a rough sense of where the limit lies in human water consumption. Where there is agreement is on the effect of this steady drain on the stuff of life. Water is running out in places where millions of us live, and people will suffer from thirst, disease and ruined economies. Worse will come, as peak water will not be measured in abstract numbers but in the register of birth and deaths.

What follows is the story of how we got into this situation. It is a long tale, stretching back to before civilisation, and then covering some of mankind's greatest achievements. To understand why there is a modern water crisis, it is necessary to understand that our chosen form of survival, civilisation, is built on watery foundations.

CHAPTER TWO

The Font of Civilisation

'THE HAND OF THE LORD came upon him. And Elisha said: make this valley full of ditches. Ye shall not see wind; neither shall ye see rain, yet that valley shall be filled with water.' This is how the Bible describes the founding of the world's first city, Jericho. It all sounds very impressive but poses a problem. The ditches must be water channels, yet there is no evidence of irrigation around the site of the ancient city. So why does the Bible describe God bestowing irrigation on the people, when there was none? The authors of the Book of Kings, writing over 8,000 years after Jericho had fallen to dust, equated success, and urban living, with controlled water. But why?

To explain how the age of water began, we need to go back to the ice. For long periods of time, Earth is very cold for humans. This is because of our orbit around the Sun; we travel in an ellipse, the shape of a gently squashed hula hoop. Various factors mean the journey is never quite the same, and that affects the temperature of the planet. Roughly, these factors combine to give us a cycle of around 100,000 years of warmth, and 100,000 of cold, or Ice Ages. We stand on a rock that weaves through space like a drunk man, dipping and wobbling, while staying fixed on the path home. Never before, and never again, will things be as they are now, when the planet is, in human terms, at its most agreeable for billions of years.

We have been walking upright on this planet for 3.6 million years. Some historians claim we made stone tools as long ago as 2.5 million years. Others put it at less, around 1.9 million years. We discovered fire around 250,000 years ago. Our ancestors, Homo Erectus, survived for around two million years in this cycle of cold and warm, with little technology to help.

Something happened in the last Ice Age to change history forever. Mankind became clever. Around 50,000 years ago, our brains got bigger. Homo Sapiens arrived, a smarter version of Homo Erectus: the brain capable of understanding an iPod or a sonnet arrived. People like you and me, with our curious minds and agile bodies, were here.[6]

It seems an infinitely melancholy thing, that we should have had the capacity to wonder at the stars when we led such a brutal existence. We are the descendants of those who survived and bred 50,000 years ago. If we take a conservative view of man's development, we have been civilised for 0.01 per cent of our time as a two-legged beast, and for only about 10 per cent of our time as a clever animal. If we were intellectually ready for the modern world 50,000 years ago, why did it take over 40,000 years to start making it?

The last Ice Age began to thaw around 20,000 years ago. Don't think in terms of a sudden warming; it took around 8,000 years before Earth was close to today's temperature, and even then cold snaps still occur. Before that, our ancestors survived in hostile conditions. They hunted for food and sheltered from the cold. Life was about survival.

Mankind's big break comes when things get warm. In other words, upright and sentient man gets lucky. A smart brain hits a good climate. It may constitute a fraction of the time we have been here, but the last 10,000 years are our rich time; we evolved and nature became more pliable to our imaginative brains. Had the climate stayed cold, we might have stayed tribal and uncivilised. Civilisation may just be our chosen way of coping with current circumstances.

We can chart the slow journey to civilisation from archaeological sites. Around 25,000 BC, humans were staring at the cave walls of Pech Merle in France and imagining the hunt of the day before, or the day to come. On the stone surface are pictures of bison and deer. Men with spears and stone arrowheads would hunt and kill them. That they had time, and safety, to portray this suggests times were good. We can conjure in our mind's eye the wall, images dancing in the flickering light of an animal fat-burning lamp, the people smiling at their achievement.

Around 20,000 BC in the warmer climate of what is now Israel, people lived in dwellings made of wood and collected grass seed, the forerunners of wheat and barley, in baskets. Five thousand years later, flint-based tools were in use.

Move forward again in time, about 2,000 years, and the grass seeds are semi-domesticated, and animal husbandry has started. With fire and farming, man edges ever closer to the brink of urban organisation. One archaeologist has called this 'the point of no-turning-back'.[7]

A shock dip in global temperature and a drought, known as the Younger Dryas period, brought a stop to these early experiments in settled living, but by 9,600 BC, global temperatures rise again, by seven degrees centigrade in a few years. This allows farming to return and become widespread. This isn't just happening in the Middle East. In the Far East, Central America and Andean South America crops are domesticated around this time. What does happen in the land to the east of the Mediterranean is the emergence of the large settlement.

Jericho was a dense pile of buildings all contained in a small space. To the modern eye it would look like a single small city block constructed without planning. This was mankind's first city. Two thousand years later, in 6,000 BC, in what is modern Turkey, Catal Huyk would be the largest settlement to date. Both relied on local springs for their water, and so were limited in size. Jericho was a mere four acres, while Catal Huyk was only one twentieth of a square mile.

As noted, neither had irrigation. They didn't need to control water in order to grow food. It would seem that this lack of ditches, despite what the Bible says, is what stopped either becoming the birth place of mankind's greatest invention. Civilisation starts in a ditch. It is the muddy channel that marks the great leap from the proto-cities of Jericho and Catal Huyk to the birth of civilisation. This occurs further south, in what is modern day southern Iraq.

The land south of Baghdad doesn't look like a valley filled with water. Criss-crossed by tank tracks, oil seems to be the business of this disputed land. The mighty rivers of the Euphrates and the Tigris run nearby, but the desert floor is hot and dry. Yet this is where the ancient cities of Ur and Uruk and Lagash and Eridu bloomed. This is where the fantastic experiment in social order and human achievement called civilisation began.

This place is called Mesopotamia by archaeologists. The word means 'between two rivers'. The people who lived here were the Sumerians. Six thousand years ago there was water across this land. The Tigris and Euphrates would flood, and water would bring life to the soil. It was good earth – it had been washed down from the mountains to the north over the previous millennia. The trouble was the flooding. When the snows of what is now modern Turkey melted, the rivers blasted water over everything, and then subsided, leaving a wet bog.

The peak flow of the Tigris was in April, but the Euphrates didn't reach its highest level until May, just as the crops were due to be harvested. Farming in Mesopotamia was a high-risk activity. The muddy waters that made the land fertile could also wipe it clean. The brilliance of the Sumerians is that they learned how to hold back the May floods, while storing enough water for the hot dry summer to come.

Dykes on the river banks, with channels directing water into small reservoirs, meant the floods didn't ruin the fields. Instead water would be directed around the land in narrow channels, when the sluice gates on the reservoirs were opened. If water could be relied on to come, then farmers could reasonably expect each crop to bloom. If people could make plans based on dependable harvests, they needn't hunt or gather. Not only could they settle, as they had been doing for millennia, but they could expect to be in the same place for a long time.

There is a vigorous debate amongst historians and archaeologists about the exact sequence of civilisation's birth. Did the big settlement come before or after the ditch? Like so much of the ancient past, we can never know for sure. All are agreed, however, a special alchemy occurred in Mesopotamia; a mixing of farming, irrigation and settlement, which gave rise to greatness[8].

In the time between roughly 3,600 BC and 3,200 BC, Ur had grown from a small settlement to a city of 50,000. Jericho and Catal Hyuk would only have been a few thousand strong. This massive growth occurred because the rulers were able to promise food to the people. This was a reasonable contract, as the irrigation ditches made grain production possible on a scale previously unimagined[9].

Keeping irrigation ditches open and regulating water flow from one part or another was time consuming and labour intensive, so a lot of people were required. These workers were rewarded with a steady food supply. Not needing to hunt or forage for meals, they settled down. An important thing happens when humans stop moving from place to place in search of water, food and safety. They have more children. Babies are a burden if you are constantly on the move, threatening your speed of action and safety. If you settle down, the value of children increases, as they provide more hands for labour. Thus irrigation creates a virtuous circle for population growth: more babies mean more mouths to feed, but also more muscle to do the work, so the irrigation ditches can be extended and more food produced, and so on.

Eridu and the Ur may have begun as tribal settlements – a few people of the same family or tribe. When the numbers increased, we see the emergence of a co-operative system. Irrigation demands co-operation.

The water flow begins at one point, higher than the fields themselves. A sluice gate is opened and the water runs into the ditch. The temptation for the owner of the first field is to take as much water as possible, but this will deprive those in the lower fields. So people learn to take their fair share and then let the water carry on down the ditches to others.

From this basic form of agreement develops a more sophisticated political structure. Archaeologists reckon that irrigated land begins in common ownership. It soon leads to private holding. This is because, as the settlement grows, more people become involved in non-farming activities. While in small numbers, the food may be shared with these non-farmers, when the population increases, they are required to buy their grain and bread.

Some farmers become more powerful than others – they exercise this power by making the poor into labourers, with no free share of the crops. Ur is ruled by a royal elite, supported by a priesthood. The king holds the power, of water and food, and the priests describe this as a divine right. Unlike tribal society, where power is conferred on the strongest and fittest, this governing class gain their authority from controlling the water supply.

In return for military protection and a steady food supply from the leaders, the poor will labour in the fields. However for this to work, the leaders need to develop a method of organising the people on a daily basis. So bureaucracy is born. We know from the archaeological records that the hierarchy was sophisticated, with many different levels of command. Officers were charged with monitoring the water supply in the irrigation ditches and then counting the yield from the fields. The bureaucrat starts life not in a suit with a clipboard, but in a ditch, regulating water flow.

The bureaucrat's art was also needed for a crucial element in this system, the storage and transportation of grain. Storehouses full of last season's crop were insurance against famine and the fates. However, to feed the people the food needed to be transported. Much is made of irrigation as a cause of civilisation, but we shouldn't overlook the watery channels as a means of transportation. Mesopotamia is where mankind first escapes the limitations of carrying something on an animal's back. By loading barges with grain, this is the first society to be able to distribute food over a wide area, and to be able to trade in surplus food stocks. People are no longer dependent on the local harvest for all their food. 'It means any shortfall in production can be over-ridden' says the archaeologist Tony Wilkinson of Durham University[10].

The bureaucrats needed a way of recording the flow of water in the ditches, and the quantity of grain in the storehouse. Writing was invented around 3,300 BC; sharp reeds were used to scratch symbols on wet clay. Earlier tribal societies may have had some form of basic marks to communicate, witness the cave paintings at Pech Merle, but in Sumer the hieroglyph or pictogram becomes hugely advanced. The records were kept on clay tablets, or papyrus. The discovery of libraries of tablets helps us understand this ancient society.

Sumerian had up to 2,000 separate symbols. When this language is revived, a thousand years later, by the Akkadian people, the hieroglyphs are converted into an alphabet[11]. Of this alphabet, the first symbol was 'a'. It was a representation of the hieroglyph for the root substance upon which civilisation was built. It stood for 'water'. The Greek 'alpha' and the modern 'a' are derived from this.

Persian is the language most closely related to Sumerian and Akkadian. In it, the first letter is 'ab', which means water. From this, the word 'abad' is formed, which translates to English as abode. So, water is home, but it is also a built home. This is important, as it signals how the Persian word 'abadan' comes about; it means 'civilised'.

Cuneiform tablets record the running of the temples. They show us that life was good in Mesopotamia. In the small city of Lagash, a tablet records that there were 48 bakers and 31 brewers. Also there were slaves, cloth workers and a blacksmith. We learn that plots of land were being rented, and the hours people worked. The distant past can seem unfathomable, but the picture that emerges from the written records is familiar to the modern reader. A pattern has been set which endures over millennia.

The water societies of Sumer developed the first legal codes and the first recorded religions. The building at the centre of Ur is a temple. It stands above the muddy soil on a small rise. This temple appears to have conferred on the ruling élite legitimacy from the gods. At the heart of this building was a pool of water.

If this all sounds a bit miraculous, or simplistic, then here is the pinch of salt. Professor Edgar Peltenburg excavated a site at Jerablus in Syria. It revealed that dry, or non-irrigated, farming settlements developed writing, urban development and monumental architecture. Northern Syria had enough rain never to need irrigation. If these are signs of civilisation, then you have to say water control isn't essential for creating the civilised world. Yet Professor Peltenburg is in no doubt 'that a kind of magic occurs with water. Settlements with irrigation, and canal transport, take off and get much bigger than other ones. They become far more interestingly complex. Out of that complexity, the qualities of civilisation develop far beyond what you get elsewhere.' He suggests water was the ingredient that made the difference between short term experiments in civilised living, and the long flow of civilisation itself. Peltenburg marvels at how, 'if you are up for it, and can cope with the consequences of water control, it will take you very far'[12].

By 2,500 BC the Sumerians had cities that were fully plumbed. Aqueducts brought water to supply fountains, baths and latrines, and sewage pipes took dirty water away – some Sumerian homes had flush toilets. In the fields, the once temperamental rivers, which would flood and destroy, were tamed into irrigation canals and dykes. The German hydrologist Gunther Garbrecht says 'The hydrological chaos of the valleys was transformed into flourishing gardens, fields and meadows that in mythology was named the Garden of Eden.'[13]

The greatest thing to flourish in this 'Eden' was perhaps the city. As the historian Gwendolyn Leick puts it, 'The most remarkable innovation in Mesopotamian civilisation is urbanism. The idea of the city as a heterogeneous, complex, messy, constantly changing but ultimately viable concept for human society was a Mesopotamian invention.'

Around 3,000 BC there were eight cities in Sumer, of which Ur with a population of 50,000 was the largest. The newly formed streets, populated by a new class of human, someone not bound to hard labour or hunting, were like laboratories for civilisation, crying out with discovery and invention.

Sandra Postel is the leading figure in the history of irrigation and water control. 'The inventions and advances of this non farm class spanned metallurgy, weaving, ceramics, specialised crafts, writing, architecture, and mathematics... irrigation unleashed a profound transformation in human development, and created a new foundation from which civilisations sprung and blossomed.'

For all the glory and power of the Sumerian civilisation, it waned, and people moved north to the city of Babylon. Here, a new empire arose, of the Akkadian people, under King Hammurabi. Never again would southern Mesopotamia be successful or powerful. Civilisation left home, and did not come back.

However the process has begun, of the river of civilisation flowing on from era to era, of new people learning old tricks. Hammurabi beats the glories of Sumer with amazing water schemes. He builds a canal from Kish to the Persian Gulf, a distance of 60 km. This enables a massive expansion of irrigation, thus multiplying the number of people who can be fed.

As Babylon bloomed into the legendary city of voices and gardens, Hammurabi also granted his people law. The Code of Hammurabi has 258 regulations, some dealing with the business of water flow and irrigation. Hammurabi's mastery of water was what conferred on him his power, even if it did require armies of slave labour to achieve. He didn't look to the heavens for his authority, but as he says in the prologue to his legal code '(I am) the exalted priest... who supplied water in abundance... who helped his people in time of need.'

In time, Babylon will fall, and the river of civilisation will move on, going north to Assyria, where the people adopt the same legal code, and refine it further into something resembling a social contract – landowners are bound to co-operate with the people in the management of water for the common good. Despite the Tigris at this point being a kinder force, flooding less, and the climate being wetter, the Assyrians adopt irrigation. They could survive without it, but it has become a symbol of power and control. Thus civilisation is now in full flow, its source hidden, but the current carrying the same ideas and symbols to all.

Queen Sammu-Ramat, who ruled during the ninth century BC, had inscribed on her tomb, 'I constrained the mighty river to flow according to my will and led its water to fertilise lands that had before been barren and without inhabitants'. The queen's successor some hundred years later, King Sennacherib, would build an 80 km long canal to bring water to the capital city Nineveh. A recovered mural or vase decoration from the site shows beautiful gardens and orchards.

The last of the great Mesopotamian empires is the Sassanian, who come from the mountains, and conquer the flood plains that have nurtured civilisation's birth. Learning from the flow of history and knowledge, they too master irrigation, and are thought to have controlled an agricultural expanse 50,000 km, large, which sustained five million people.

This journey, from Sumer to the Sassanians, has taken around 4,000 years – a significant majority of the time that civilisation has been around. Travelling from Ur to Nineveh, man acquired most of the core elements of civilisation. He had gone from a field to a city of 50,000, to an empire of five million. Law, writing, politics and bureaucracy became facts of life.

That is why, when the Book of Kings was written, and Jericho had to be explained, it needed a valley full of ditches. The ditch was not only the proof that Jericho was great, like the high towers of Dubai are proof that it is as ambitious as 20th century New York, but that Jericho existed and was able to function at all because, from Mesopotamia on, the idea of the city depended on water control.

Rebuild Dubai, the one we destroyed at the beginning of Chapter One. Put back the buildings and shops and the street life, the organised chaos of the city, regulated by politics, bureaucracy and law. In your mind, put back the noise of the modern world, the crush of friends and strangers. Throw in the issues of the age, about climate change and quality of life and globalisation; all this is rooted in the simple ability to channel water.

The thing we call civilisation is wet, like the blotting paper that children grow cress seeds on, and our achievements are the green growth. The question is, will we still be civilised if the water runs out?

CHAPTER THREE

Civilisations Bloomed from Water

THE CHINESE HAVE a mythical man, representing strength and power, a sort of giant who looms over the land wielding what looks like a wooden lance. The muscled figure is Yu, and it is not a weapon in his hand, but a hoe. The long-handled spade was essential for digging and clearing ditches. Mighty Yu, virile and proud, is born of the crucial technology upon which the first Chinese civilisation was built – irrigation.

Cities bloomed between the Tigris and Euphrates, but what about elsewhere? In this chapter we will look at the civilisations of China, the Indus and the Americas. The lesson is that all great experiments in human development from this point on depend on water control. While many peoples learn to dam and irrigate, only a few become successful civilisations. This has as much to do with luck as anything else: civilisation is a hit and miss affair, its grip on the landscape as strong as mud.

We have travelled thousands of miles from Mesopotamia to the Yellow River. It stretches down from the Bayan Har mountains to the coast, nearly 2,000 km as the crow flies, but twice that in length as it twists and turns through northern China. Like the Yangtze, it runs from west to east, cutting across the nation, linking people vast distances from one another. It is one reason why a nation so large can exist – the rivers have kept the people connected.

The rivers have also blessed the land with the force of civilisation. The business of controlling these snakes of water, capable of twisting violently against man's needs, united the people and gave them a political structure. The water also gave them the resource to grow crops enough to feed a large population, and the power to run silk mills and metal workshops. To find the answers for China's power, and for its relative isolation from the rest of the world, you need to look into its currents.

Archaeological remains suggest dykes were being used to channel water around 6,000 BC. Remains of bamboo piping indicate the Yellow River civilisation had rudimentary plumbing as early as 4,000 BC. We do know that a large, organised society existed around the flow, digging irrigation systems and feeding many thousands.

Yu's mythical strength may have inspired or intimidated the people of the early Chinese civilisation, but they certainly knew that the gift of irrigated land was won at a terrible cost. The Yellow River carries a great deal of silt. Silty water is rich in nutrients, washed down from the mountains above, and good for growing crops. The flood plain of the river was a land that promised bounty. However, large rivers carrying a lot of silt are also, by natural design, given to changing course. When the water slows down at any point, the silt drops to the river bed, builds up and can create an island or enough of an obstacle that the water chooses a new path. If you have spent a great deal of effort constructing irrigation ditches, you want the river to keep flowing along a predictable route, so as to keep your fields wet. The result of water's natural whims being resisted by man's need for organised ditches is that a constant battle rages between the silty river and the sweat-slicked peasant with a hoe, endlessly clearing dirt from the riverbed and the man-made channels. Yu wasn't a kind of terrible colossus, but an ideal, a figure that peasants must have wished they could be, with the strength to cope with this endless task.

In order to keep the river following the same bed, even as it deposited tonnes of silt, the Chinese needed to keep raising the height of the banks. As the riverbed got higher from silt, so the dykes holding it on its path got higher. In modern times the river is about six metres above the land. In the last 100 years disastrous floods have occurred. It is reasonable to assume that flooding must have been a problem from beginning. Though not intended as such, Yu was also a symbol of a bittersweet life, of the never-ending need to keep digging.

A worker in the vast plain of the Yellow River might literally look up to Yu, this exaggerated man of water, but would also spend a life in his shadow, because Yu represented political power. By 1750 BC, to control water was to control the fate of an empire with cities. Vast numbers were dependent on irrigated agriculture, and these people needed to co-operate if they were to manage the restless waters. Only a few individuals held political power, while most able-bodied people held a tool like Yu's and spent an exhausting life mastering the flow of the Yellow River. In Mandarin the word 'schin' means both 'to regulate water' and 'to rule'.

Chinese civilations rose and fell, but by 1,000 AD, the Yellow River fed 100 million people. It was the most advanced culture in the world. The great achievement was a canal linking the Yellow River and the Yangtze. It was 1,800 km long, and kept flowing by a series of reservoirs and the new invention, lock gates.

The Chinese created an amazing civilisation, but glorious as it was, it didn't set the model of what being civilised meant for the rest of the world. In this, it is not alone.

The Indus River, which runs from Tibet to the Indian Ocean, helped on by tributaries spanning from Kashmir to Afghanistan, sustained the most remarkable civilisation. Five thousand years ago a string of a thousand towns lined the banks of the river. Every settlement was made of mud bricks of the same size, and every town followed the same model, to make what the writer Alice Albinia calls 'probably the world's first planned cities'.

Archaeologists refer to this lost wonder as the Indus Valley Civilisation. Our knowledge is limited as the language remains undeciphered, but artefacts suggest this was superior to anything in Egypt or Mesopotamia.

Wooden ships plied the river, supplying food along the length of this civilisation. Large irrigation schemes, using brick dams, collected the wealth of the water and spread it on the fields, where many crops were grown. For the first time in human history, cotton was domesticated showing they were an extremely sucessful society. This would become a major source of income for the empire, as the white cloth was sold in Egypt and Mesopotamia from small trading settlements established in these regions. Later, the farming of cotton will have a dramatic effect on this area.

Mohenjo-daro and Harappa are the two largest surviving ruins. Both have suffered from chronic thievery – not least subsequent empires and religions stealing the bricks to build new monuments and railways. Enough is left to show cities with clear street plans, solid two-storey town houses, internal bathrooms and sewage pipes which connected to an urban waste system. It would take Rome another 2,000 years to match this sophistication, and northern Europe would need 3,500 years. Unlike Mesopotamia, these cities are not built around temples, and the grandest buildings are not palaces. At the centre of the towns we find grain stores and public bath-houses.

Water control is only needed in places where the supply is variable. The crucible for civilisation is struggle; it's when nature has dealt man a tough hand that he is forced to respond by collaborating with others and developing technology. If the circumstances are too difficult, then man reckons bare survival is the most he can hope for. If things are easy, then man has no need to bother.

The Americas posed any number of challenges to survival. To the west of the Andes, the strip of land before the sea is narrow. Water rushes down the mountains and speeds past the soil. Farming is hard. On the mountains, steep slopes make farming harder, and altitude limits the choice of crop. To the north in what is now Mexico, settlements were endlessly caught in tribal wars, while in the Yucatán Peninsula to the eastern coast of Mexico, an unpredictable climate made for a fragile existence. It was the difficulty of living in these places that made them home to several communities.

Allowing for the vigorous debate that continues among archaeologists as to when man first arrived in the Americas, it seems reasonably safe to assume this was over 30,000 years ago. The oldest site yet excavated is at Monte Verde in Peru. Evidence here suggests a settled population over 12,000 years ago, with some artefacts suggesting human activity around 32,000 years ago. Either way, it is beyond dispute that America was developing social communities long before Europe had a hut. 'If Monte Verde is correct, as most believe, people were thriving from Alaska to Chile while much of northern Europe was still empty of mankind and its works' says the historian Charles Mann.

Again, there is much academic debate about exact times, but a majority view now argues that only Sumer beat the people of what is now Peru to the prize of civilised living. It is possible that subsequent research will show the culture of Norte Chico, on the Peruvian coast, predates Ur and the glories of Mesopotamia.

The people of Norte Chico gathered at the confluence of four rivers – similar to the other early civilisations. But here there is a crucial ambiguity. Nestled on the narrow strip of land between the sky-scraping Andes and the cold swell of the Pacific Ocean, they weren't looking at the large flood plains of Egypt, Mesopotamia or China. They had a small stretch of land, but to one side the sea, and the other the hills.

They had irrigation, but it would seem this was not for the mass food production that occurs in other civilisations: Norte Chico was different because the land was different. It wasn't a question of making do with one type of environment, as in Sumer, and harnessing that to maximum effect. The early Peruvians have three environments to make their living from – highland, lowland and saltwater. By using all three, they could produce enough food for their people without turning irrigation into a mass-participation necessity.

This may be important. When the Europeans did come, every American civilisation collapsed, and with speed. One reason may be they lacked the political cohesion of mass-participation societies. It is certainly true that while American civilisations relied on water control to establish themselves, they evolved in different ways to those in the Middle East or China. Perhaps they were different because they had a choice.

We know from archaeological study that cotton was grown in the irrigated fields of Norte Chico. You can't eat cotton, but can dress yourself in comfortable, breathable layers. To give land over to a non-edible crop suggests that enough food was coming from other sources. The cotton was not just for clothes, but also for making fishing nets. These people exploited the sea, the fast water from the four rivers, and the hunting possibilities from higher land. Irrigation was there, but not as a defining technology. This does not set a pattern for civilisations in the Americas, but does hint at a recurring theme – water control is fundamental to the establishment of civilisation, but never becomes the central, unifying task it is in the Euro-Asian societies.

There were two great areas of human development in the Americas. These are the Andes (Peru) and Mesoamerica (Mexico and the central American nations). It can seem that Peru's sole claim on the modern global imagination is the amazing city of Machu Picchu, perched on a mountain like the nest of a brilliant bird. In fact, within the borders of the modern nation, and stretching down the ridge of the Andes to Chile, is arguably the richest area for human endeavour outside the Fertile Crescent east of the Mediterranean. It is a treasure chest of humanity, revealing endless mementoes about our collective past.

A long chain of settlements led to the amazing achievement of the Incas. Considerably later than Norte Chico, we find the Chavin civilisation, dating from around 800 BC to maybe 200 BC. Further south the civilisation of Tiwanaku emerges around 100 AD. On the shore of Lake Titicaca, the people drained land and planted crops to feed a large city. A thousand years later, the population is estimated to have been over 100,000 people. While Rome fell and Europe dragged itself out of the bog, Tiwanaku was a grand metropolis, with pyramids and monuments.

Running water was supplied to the citizenry, and the waste taken away in closed sewers. Historians believe the later Inca society was so efficient at irrigated food production, they had eliminated hunger from society, and certainly European invaders would marvel at the excess grain in storage[14].

In coastal Peru, the Chimu Empire emerged. Its time of high achievement comes between 1200 AD and 1500 AD, but evidence suggests it emerged out of irrigated communities around 1,000 BC. Its capital city, Chan Chan, was a bustling one of 50,000. People lived within a centralised bureaucracy, amid temples and pyramids.

Go inland and we find the Beni civilisation. It appears to be a perfect example of how large populations emerge from water control. Begun around 3,000 years ago, this society grew by learning how to control the seasonal rains on the grassland. With houses and temples built on high mounds, the entire area would be flooded, and the fish from the swollen rivers caught in elaborate systems of channels and nets. They had canals and dykes and reservoirs, and from this stability grew a population estimated at a million strong by 1,000 AD. Their food production was so successful that some of the irrigated land was given over to cotton.

In Mesoamerica a similar pattern of civilisations emerge, flourishing then withering away. The Olmec are the first civilisation of Mesoamerica, beginning around 2,000 BC. They settle in the Tehuacan, south-east of the current Mexico City, on a flood plain, not dissimilar to that around the Nile. They develop an elaborate irrigation scheme. They have a written language, vigorous trade routes and a recorded history. Roughly 2,000 years later, in the basin of Mexico, around Lake Texcoco, the civilisation of Teotihuacan emerges. Though it constructed the third biggest pyramid in the world, and has left extensive remnants of a city, we don't know that much of this culture as the language has yet to be deciphered. However, one can tell from the architectural remains that they had reservoirs and running water through the city.

A feature of the Mesoamerican civilisations is the process of dredging boggy land to create a landscape of canals for transport and mounds of fertile soil. Known as 'floating gardens', this effective way of reclaiming bog doesn't occur elsewhere, and may also explain why the Mesoamericans had little need for draft animals, or wheeled carts[15].

Further south the astonishing civilisation of the Maya was founded in the Yucatan peninsula. Here the people found ingenious ways of storing water in the small depressions of the land. The city of Takal for example had a storage capacity to meet the water needs of 10,000 people for over a year. From this stability, they built huge temples and cities, and a vibrant intellectual culture.

While these civilisations developed the same irrigation techniques, and mastered their water supply in relation to the environment, they didn't follow the same path of growth as in Mesopotamia, China or the Indus. The political structures that arose were so rigid that when the leader fell, the society was at a loss. They did invent, and at roughly the same time, language, religion and bureaucracy. In their time they were the match for, and superior to, rivals elsewhere in the world but one has to be very sensitive to South American tradition before you can spot any remains of it in modern global civilisation.

The Yellow River, the Indus Valley and the various civilisations of the Americas all have elements of what occurs in Mesopotamia. A time traveller visiting each at their peak would struggle to spot the ones that would survive and those that would fail. While we can comb through history looking for clues, we shouldn't miss the obvious point. Survival may be as much about luck as anything else. One place was particularly lucky, and it created a civilisation that would last for longer than any other, and would become a beacon of man's ability and ambition to the world – Egypt.

CHAPTER FOUR

Egypt

TO UNDERSTAND THE POTENCY of Egypt, and by association water control, you should imagine it like a combination of 20th century USA, modern Dubai and Chinese economic potential all in one. Here was the strongest economy in the world, the richest society and the most desirable lifestyle. This glorious flare of luxury and power burned so brightly because of its ability to master the Nile. For much of man's civilised history, Egypt and the Nile has been the standard that others yearned to match. It is only comparatively recently that Greek or Roman civilisation has usurped Egypt as the supposed font of Western civilisation.

If civilisation is a river, we tend to think of Mesopotamia as the source, and imagine it moving downstream to Athens and Rome. The emergence of a strong and enduring European 'civilisation' in the 15th and 16th centuries AD is described as a Renaissance, a rebirth, of what happened in Athens and Rome. In the popular imagination, Egypt gets left to Elizabeth Taylor as Cleopatra and holidays to the pyramids. This is wrong. To understand the stream of civilisation, we need to appreciate the magic of the Nile[16].

Magic is appropriate. The Nile behaved like no other river. While the Yellow River constantly threatened to breaks its dykes and the Indus and Euphrates could prove fickle sources for large irrigation schemes, the Nile was different. Its cycle of low water and high, its great dump of rich silt, was more generous than any other riverbed. Other currents raged and ebbed, but the Nile was like a woman's cycle, regular and fertile. Indeed, like an impregnated womb, it rose magnificently, in a steady way, signifying the bounty that was to come. From this unfathomable kindness of nature grew and survived the longest-running civilisation known to man, resplendent with astonishing monuments, great wealth and intellectual richness.

Around 15,000 BC, a verdant land of trees and grasses spread over a wide area of Egypt. For climate reasons, this benign environment steadily began to shrink, letting the Sahara Desert push the plants and humans back to the Nile. Around 7,000 BC there was a slither of green clinging to the edges of the great river, sheltering humans who had retreated from the relentless waves of hot sand. If a folk memory of what they had lost over the millennia existed, it is unlikely to have made them sad, for they were clustered on the banks of a natural wonder.

The Blue Nile stretches down over hundreds of miles from its source in the Ethiopian highlands. These hilly lands are green and rich in plant life. Here the river swells with melted snow in July, scraping rich soil off the land. These waters meet the White Nile at Khartoum to become the Nile proper. Two months later the high waters will have reached Alexandria on the Mediterranean coast, having flooded the plain up to one and a half metres in depth. Weeks later the flood subsides, dumping a black mud of rich mountain earth. The river returns to its bed and the land is primed with nutrients, ready to nourish crops of grain. On this reliable supply of food, the power of Egypt was built.

For Egypt alone, the flood is a good thing. The high waters don't represent divine wrath and a bid to clean the land of sin, as in other religious myths, but are proof of celestial bounty. Unlike the thousands who laboured, yearning for Yu's muscles, on the banks of the Yellow River, the people of Egypt had no choice but to stand back and let the flooding water find its course.

The Nile deluges the land. People survived the swelling of the river by, at first, backing-off and letting the water do its wish, and then by constructing an irrigation system that allowed the bounty of the delta to be spread further than nature intended.

In the first bloom of Egyptian glory, farmers didn't bother with irrigation, relying on the fecund swell of the river. When, for reasons that are not entirely clear, the river's floods were markedly reduced for a period of several hundred years around 3,500 BC, Egyptian civilisation teetered on calamity. By the time the high waters returned, the people had learnt their lesson. Egyptians became much better at directing water away from the riverbed and into channels.

This distinction in water technology is important. The Sumerian civilisation rose and withered within the space of a thousand years. This set a familiar pattern for civilisations. Few lasted for more than a few hundred years. While diverting the Tigris or the Euphrates could be the key to a power, it was also, frequently, the path to ruination. The sheer work of maintaining large irrigation ditches was a drain on the empire, while the risk of the land becoming contaminated with salt, thus ruining fertility, was high. In the period between the start of records, 3,300 BC, to the invasion of Alexander in 330 BC there were more than three dozen civilisations in the Near East region.

There would be no need to spell out the dangers to a labourer working in the shadow of Yu or on the Indus Plain. Controlling water was a devil's contract, where the benefit of food was paid for with huge expense, loss of life and potential collapse of the civilisation. Egypt's greater reliance on natural flooding explains why it thrived for so long. The Egyptians never had the opportunity to make the poor choices that condemned other civilisations.

The river delivered an annual bounty, but that doesn't make agriculture or civilisation inevitable. The key to Egypt's success is the scale of the irrigation task. The swollen river still made mankind work for his benefit. The business of controlling these waters was beyond the strength or wit of an individual. To make the water serve farming, large earth works needed to be constructed. This meant co-operation was essential. The glory of Egypt was partly a result of the Nile, but just as importantly was the combined effort of a large workforce, all labouring towards a single end. It was this creation of an army of muscle that would allow the pyramids and other wonders to be built.

People further south, with just the same river and just the same rich waters, didn't collaborate. They stuck with primitive farming methods. The consequence was that no great civilisation arose on other sections of the Nile.

Egyptians first used basin irrigation. After the bad years of around 3,500 BC, this new system allowed farmers to capture floodwater in small reservoirs. The water would be allowed to stand, dropping some of its silt, and then the basin would be breached, flowing into another basin, and so on. Perhaps this is what is depicted on the ceremonial mace held at the Ashmolean Museum in Oxford, England. Inscribed in this stone is an image of the Scorpion King digging a trench to a date tree with a hoe. This dates from 3,100 BC.

Egyptian farmers invented tools such as the *shaduf*, the Archimedean screw and the Persian wheel to lift water from the river. These simple devices collect water and direct it towards fields, orchards and gardens. They turned the Nile valley into a breadbasket, which sustained a population of 180 people per square kilometre[17]. The historian Sandra Postel says that by the time Egypt was providing the Roman Empire with wheat, it had one million hectares under cultivation.

In Egyptian myth, the father of all gods is the god of the Nile, Hapi, and people worshipped this deity with hymns.

> 'Praise to you, O Nile, that issues from the Earth, and comes to nourish Egypt...
>
> If his flood is low, breath fails, and all people are impoverished; the offerings to the gods are diminished, and millions of people perish. The whole land is in terror...
>
> When he rises, the land is in exultation and everybody is in joy...
>
> He fills the storehouses, and makes wide the granaries; he gives things to the poor.'[18]

Hapi is a man, but with hermaphrodite qualities. Egyptians thought of the river as male, with the possibility they considered the flow like some kind of ejaculation of semen. Equally, the riverbed itself can appear like a female figure, impregnated by the flood.

At Cairo and Elephantine there were Nileometers, which measured the height of the river. There was a precise gauge of the happiness of the people from the swell of the river. Sixteen cubits was the ideal – it meant a summer flood of just the right amount of water. Fifteen cubits meant security, that is, enough food for all to be fine. But 13 cubits meant hunger, and 12 represented famine.

Around end of the third millennium BC the Nile dried up. Twelve cubits was the measure of the river, and famine was the state of the nation. Egyptian civilisation sank into the sand in a very short time. Like all 'civilised' populations, when the water supply is insufficient for food production, the core justification of the community is gone.

Archaeologists have uncovered a literature of what is called 'lamentation' which describes the terrible plight this golden empire descended into in a short time. With food low, the economy imploded and people committed suicide. Others tore up the civilising elements of law and social practice and anarchy ruled, most profoundly in a return to cannibalism. In mythological terms, this is portrayed as a battle between the god Osiris, who is 'civilisation' and his nemesis Typhon as barbarism.

Low water wasn't the only threat. If the summer swell took the river above 16 cubits then food stocks, seed stores and temples would be flooded, and this could be equally dangerous, as it jeopardised the whole economy. Historians refer to the organisation of pharaonic Egypt as a 'temple economy'. Similar to Sumerian society, the grain was collected and distributed by the priests. The temple was both a place of worship to the gods, and a warehouse of plenty. The absolute link between religion and eating created a loyalty to the most senior religious figure, the Pharaoh.

Egyptian priests would honour the river at libation tables. These were small stone surfaces that were engraved with pastoral scenes and a maze or pattern of channels. When wine was poured on, it would run through the chiselled courses like water through the river bed, past images of corn and grapes and the great animals of creation, lions, crocodiles and dolphins. Alternatively, presumably presaging some tragedy, the priest could inundate the table such that wine flooded everything, but was still contained with the meandering pattern at the border.

This isn't a book about Egypt, which alone could account for volumes on the subject of water and civilisation, so we must follow the current on. As the journey continues, the collective imagination, or the folk memory, will repeatedly call back to the pyramids and obelisks of this land. If Mesopotamia did so much to foreshadow civilisation's path, then Egypt lit the way.

What happens next is that Mankind begins to lose his connection to the river. If the triumph of the early empires is that they tame unpredictable flows, from about 1,000 BC on, civilisation begins to take the control of water for granted. Something which had been magical becomes a given. It is still the source of power, and the determinant to an empire's success, but the gradual process of burying water under the complexity of the state begins. We lose our immediate sense of water's importance. From now on, man will no longer follow the water, but get the water to follow him[19].

CHAPTER FIVE

Athens and Rome

PERHAPS IT IS NO ACCIDENT that this story meanders. The word is derived from the River Maiandros, which takes a lazy slalom through Greece. So the course of civilisation is not straight. It drifts in one direction then the next. Of course there are many factors influencing it, not just water control. Indeed, one of the key issues is how the necessity of controlling water gets forgotten amid other concerns. Perhaps the first major instance of this is in the Greek and Roman civilisations.

Athens and Rome have a different attitude to water than Ur or Alexandria. In Sumer or Egypt, the irrigated fields were visible from the walls of the city. This direct link between irrigated land and the city is broken in Greece. For the first time in human history, people don't live in walking distance of all the water they need.

The Greeks rely on food coming from elsewhere. Instead of creating a symbiotic relationship between irrigated field and city, they literally farm out the business of food production, and pay for it with money created via military or economic means. Compared to any previous civilisation, the grain to make the morning bread was coming from huge distances away, in this case the island of Euboea or from around the Black Sea. Like any settlement, they needed their food, but they didn't need to control water for its local production.

This may seem a minor adjustment. It also seems obvious – the modern urban world is far removed from its agricultural support. Yet the idea of this separation first occurs in Athens, and with it an important shift in thinking. A Sumerian would have no hesitation in telling you that the city was about water. An Athenian would have told you the city was about people, democracy, or economics. Though the river of civilisation runs on, the water becomes less visible.

The Greeks concentrate on defining the idea of the city. Political structure and personal responsibility come to the fore. We are familiar with this, as the western world inherited and still applies many of these ideas, though much changed over time. The Greeks give us the word civilisation, from the root 'civis' meaning the people. In other words, the bundle of stuff we may mean by 'civilisation', such as law and bureaucracy and political order, is credited to the control of water up until Athens. After that, it gets credited to human ingenuity. Water's child, the city, has left home and invented a new life story.

This doesn't mean the Athenians don't rate water control. It is still an important technology, and very much associated with civilised living. It is just no longer the root of it all. Instead, controlled water becomes a signifier of civilisation; in particular, to be clean is good, to be unwashed is the opposite of civilisation, or barbarian.

Fresh spring water bubbled and rushed into Athens, directed to the central fountain in the market place by sophisticated aqueducts. The favoured method was to run the water through clay or stone pipes that were laid on the ground. Some stretched for 40 miles, and ingenious siphon systems kept the water running, even up hill.

The grandest example of Greek siphon technology is at Pergamon. Three lead pipes run side-by-side down and up a valley just over three kilometres in width. The Madradag aqueduct carried a volume of 4,000 cubic metres of water a day[20]. Where was all this water rushing, if not to the fields?

The importance of the Madradag flow and its daily delivery of mountain water was felt on the tongues of the free citizens of the Greek cities. They drank it, used it as medicine and washed in it. The public fountain in the Agora, the central market place of Athens, had basins 20 feet by 10 feet wide. Water also spouted out from holes in the wall, so that a jug could be filled directly.

The wealthier would also indulge in a bath, in water supplied from far beyond the city walls. These are not the lavish bathhouses of the Roman Empire, but rather practical ones. You would sit down in a shallow pool of water, enough to cover your bottom, while your feet were in a sunken pool. Thus you washed, as if in a chair. The very rich fitted bathrooms in their houses.

Thankfully the Agora also had a large sewer, a stone-covered channel that drained into the Eridanus Stream. As there is no evidence of public toilets in Athens, these would have done something to diminish the smell of city life.

To go to the considerable effort of building long aqueducts, which must have been expensive, indicates the importance of clean water to the

Athenian sense of civilised living. Life was possible because mountain water endlessly rushed into the city. While the wealthy of Ur or Memphis no doubt washed, their water supply was a by-product of agricultural need. For Athens, the water was worth the effort because it was a sign of how civilised they were – washing alone justified the expense. The quality of civilisation had developed – cleanliness had become a goal worthy of all civilisation's effort.

Rome needed Africa like Ur needed its fields. In Algeria there is a stone that testifies to the link between the empire on the northern coast of the Mediterranean and the lands on the southern rim of the salty water. At Lamasba there is an inscription which gives the schedule when each plot of land should be irrigated and how much water it should get. The instructions were for local labour, but the food was intended for elsewhere. From what is now Morocco to Egypt itself, African land was Rome's larder.

The link between food production and the city, the literal and necessary proximity that saw the concept of civilisation emerge, was broken. Rome contracted out its wheat and barley growing to distant lands – a process we might now call globalisation had begun. Military power, economic strength and superior technology made this possible. Had Rome relied on its local agriculture, the empire would never have triumphed.

There is no evidence of mechanical irrigation in ancient Italy. That is, the use of *shadufs*, dams, ditches or water wheels. The agriculture of the region is described as 'dry farming'. The leading historian on the Roman Empire's water use, A. Trevor Hodge, says farms 'were never far from total failure'[21]. This was not a civilisation built on local irrigation.

Writers such as Pliny and Virgil do describe irrigated vineyards, crops and gardens, but the technology was not appropriate for widespread use. This seems strange – when one knows how important water control is to the invention and development of civilisation, for the most powerful empire before the modern age not to have that umbilical link to the ditch rattles the mind.

It also appears ironic. Rome was begun in 600 BC, built on the banks of the Tiber where the river's flood plain narrowed. This was the lowest bridging point. The first span across the waters was the Pons Sublicus, constructed from wood. As the bridge was so important, being in charge of it, even notionally, was the most prestigious of jobs. So the chief priest of pagan Rome was called 'Pontifex Maximus' – essentially, 'bridge builder in chief'. The world still has a 'bridge builder in chief', though the holder would be horrified to be called a pagan, as the word evolved into Pontiff, otherwise known as the Pope.

Water was at the heart of the social contract. The provision of free water to the citizenry defined the relationship between the city fathers and the people. If they couldn't deliver water, then the rulers had no legal basis for their authority. What occurs in Rome is the idea that services justify political structures. It is the principle at the heart of modern, developed societies. The provision of water was a quality of belonging to Roman civilisation. The empire's control of water in Africa and elsewhere allowed it to produce enough grain to sustain a subsidised bread programme, to keep the poor fed.

However, it is the magnificent aqueducts that speak the loudest about Rome's water control. Contrary to popular assumption, the aqueducts weren't built to irrigate farmland. The multi-tiered engineering marvels were purely for reasons of 'civilised' living – they delivered clean water for baths. Much as the structures could have many layers, so they carried multiple messages about the quality of Rome.

They were an expression of military power, built by the army. Disciplined troops marching in straight lines may have seemed formidable to Rome's enemies, but when the same men could construct such architectural and technological wonders, the power of the empire must have been self-evident to all.

They were an expression of architectural wonder: several tiers of archways running across a valley floor, forming a solid net of stone which would catch the eye of any person. To a peasant in a hut, whose livelihood depended on the whims of natural bounty, an aqueduct must have appeared as a shocking statement of the ability of man, and Roman man in particular.

They were a bold advertisement of the technological superiority of Rome. Getting water to run along a course is both blindingly easy, as any child knows from playing on a beach, and remarkably hard. If the water runs too fast it will erode the pathway or flood, too slow and it will silt up, eventually damming itself. To get water to run at a steady pace over long distances requires planning, mathematical knowledge, accurate measurement and organised labour. Aqueducts weren't simply magnificent in themselves, they made no apology about how magnificent the empire that could build them must be.

Finally, they were very expensive to build. Only the strongest of economies could embark on such constructions. The aqueduct network didn't simply control the flow of water into Roman cities, it was evidence of Rome's belief in its superiority and efficiency.

Rome first depended on spring water for its cooking and drinking needs. Romans were also adept at collecting rainwater. Neither source was sufficient for a growing population, with increasingly luxurious demands. Between 312 BC and 229 BC 10 aqueducts were built, bringing water from the east of the city and the Anio Valley. Technical ingenuity was already evident. Aqua Tepula and Aqua Julia were built to follow the same route into town as Aqua Marcia, which meant a triple deck of aqueducts. These supplied water for drinking and bathing.

In 19 BC the first aqueduct, for purely bathing purposes Aqua Virgo, was constructed by Agrippa for his baths at Campus Martius. Seventeen years later Augustus built Aqua Alsietina so that he would have enough water to flood his lake at Naumachia, on which he staged mock-sea battles and other entertaining shows.

The last great aqueduct into the city was Aqua Triana, taking its supply from the west, and delivering it to the Via Aurelia, where a huge junction sent the water to different parts of Rome. The water rushing out from the pipes at Via Aurelia was headed partly for grain mills and for private houses, but mostly to the bath-houses. In 33 BC Emperor Agrippa carried out a census and while counting heads for tax purposes it usefully totted up other aspects of life in the city, including 170 small baths. Some 500 years later there were 856 baths, plus 11 imperial baths known as thermea. The total population, of whom many would be slaves and therefore barred from the baths, was at most one million people. Two thousand years ago places to wash in Rome were as common as coffee shops today. This degree of organisation and technological achievement would not occur again in human affairs until the 19th century, in industrial Britain.

The baths served a similar purpose to the modern coffee shop. They were places of social gathering which provided an apparently vital element of city living; to wash was to belong to Rome and its success, in the way that a double mocha signals involvement in contemporary society, a visible display of opting in. Washing brought health benefits, but bathing was deemed a luxury, not a necessity.

If we think of municipal government consisting of several departments covering education to waste collection, then the predecessor of all these is the water organisation of Rome. In charge of the whole workforce and network was the Cura Aquarum. The most famous man to hold this post was Sextus Julius Frontinus. As author of *De Aquis Urbis Romae* his words describe this magnificent system to later ages.

Written around 100 AD, it refers to 255 miles of aqueducts supplying Rome. It details the elaborate distribution system, whereby aqueduct water arrived at junctions called castellum. From these halls the main flow was divided into many separate courses, guided by tunnel, lead pipe and hollowed-out logs. Frontinus's methodical description indicates that water use was about 300 gallons per person per day. Per capita use in a developed western society today is around 100 gallons.

Frontinus describes the first urban distribution system. Think back to our stripped-down city at the beginning of this book, when we imagined the labyrinthine structure of gas pipes, electricity cables, telephone lines, communication cables, sewer pipes and water pipes – the veins and arteries of city life. Now imagine ancient Rome, and the network of water channels. This is what made the city possible.

As today, the system was vulnerable to abuse. One of Frontinus's constant anxieties was how to catch those people who broke into the pipes to steal water. A whole body of jurisprudence existed to govern the ownership and rights to water flow.

The aqueducts also establish the idea that not only is controlled water 'useful', as opposed to the apparently 'wasted' water which just runs down a hillside, but that the constant flow of water is a sign of power. The background to this is quite accidental.

Despite the many technological advances, there was one key element familiar to all modern water systems that did not exist. There was no tap to the aqueduct. Once running, it couldn't be stopped. Unburdened by our modern sensitivity to preserving resources, the Romans turned this constant flow into entertainment.

Agrippa was the first to recognise that objects in or close to water can look more animated and interesting. He decorated the many public fountains in Rome with over 300 statues. The figures were of gods and heroes, suddenly alive in the rush and babble of the flow. To the onlooker, this would have been an amazing sight. In an age with no moving image, the figures rising from the foam would have been a great excitement, appearing weightless on the froth, or dynamic with a slick wet surface. Here the never-ending supply of water was used to venerate the gods, and so reflect well on the power of the emperor. In a more theatrical use still of water, the aqueducts themselves were celebrated in a way which cast a benevolent light on the empire.

On approach to a city, an aqueduct is no more than a stream or river within artificial banks. Rome chose to terminate them in enclosed junctions called castellum, but that was not the only option. As the Empire grew richer, and to some more decadent, so the technology was displayed to greater effect. The third century AD aqueducts in what is now Turkey often ended in full public view, the water not going underground, but being made to cascade in lavish waves over stone façades, which would be decorated with figures and natural scenes. At Perge, water splashed down a façade and then along a cascade that runs the length of the town.

Imagine a modern theme park, and the water rides there, and then imagine this in the middle of a city. Remember that water was a precious commodity, and that to control it on any scale was a gift given only to the very strongest and richest. Now consider walking into Perge, and watching this display, and having any doubt about the might of the Roman Empire washed away from your mind.

Both Greece and Rome looked to enrich their culture by copying and adopting the knowledge and myth of the Nile. Like the water at Perge, Egyptian wisdom cascaded through later civilisations as a sign of their worth. Some Greeks thought that to find the source of the Nile (an occupation which would fascinate later European empires) would be to stumble on the font of all things. Where the Nile began, so did wisdom. As such, the river was thought to bear the gift of civilisation; quite a trick for a muddy flow.

The scholars of Athens thought if they were ever to answer the great philosophical questions of the age, they needed to understand the Pharaohs. Thus a kind of Grand Tour emerged, like those taken by diligent north European students in the 18th and 19th centuries AD to the wonders of Italy, in which writers and artists would go south for their betterment.

However, they could never go quite far enough south. They couldn't reach the source of the river, and without this the explanation for its remarkable behaviour remained mysterious. So the Nile was analysed and recorded not in empirical terms, but in myth, and in so doing, the

Greeks continued the ancient belief in river gods, and would pass this mythology on to western culture. Historian Simon Schama describes it thus: 'And with Seneca, Pliny, Plutarch, Strabo, and Diodorus, an entire generation of Nile literature – a rich slurry of myth, topography, and history – inaugurated the western cult of the fertile, fatal river.'

The historian Herodotus shows Greek bafflement at the river. He figured the origins of the river to be in the Ethiopian south, beyond the mountains and in the desert. As such, the Nile seemed to refute all that was known about watercourses. The rivers that poured down onto Athens began in the cool highlands and swelled into spring floods, only to settle to a low stream in summer. Not the Nile. It seemed to begin in an arid zone, and in summer time it was at its highest. And while it swelled into a massive delta at this high water point, which could break the supporting dykes on each side, it never roared a reckless path, tipped with white water, like other rivers.

In Plato's *Dialogue*, Critias described the Nile as a 'saviour' river. Unlike the destructive forces of Greek rivers, Critias tells of the Nile rising up and gently swelling over the land. He says this absence of destructive floods explains why Egypt's monuments are so well preserved. They were never ravaged by angry water rushing from broken river banks. Plato thought that because the temples and pyramids still stood the Nile valley was place of memory and the past.

Herodotus may have stepped back from the puzzle of the river's behaviour, but others waded in. Diodorus Siculus was a popular historian who wrote a 40-volume account of the known past, the *Bibliotheca Historica*. In it he adapts Egyptian myth to suit Greek purposes. He plays a trick as old as civilisation – he changes the past to suit the present.

He takes the story of Osiris and Isis and retells it to blend in with Greek myth, thus spinning a tale that appears to plant the roots of the Greek empire in the soil of the Nile. The primeval river myth for the Greeks was described in Hesiod's *Theogeny*. Achelous fought Hercules for the hand in marriage of Dienara. During this battle Achelous transforms himself first into a serpent and then into a bull. Neither incarnation was strong enough to defeat Hercules, who, during a final fight, rips off a horn from the bull's head. This is thrown into a river by nymphs and it sinks to the bottom. On the bed the horn sprouts into another form of life, one bearing fruit and food. It is the Cornucopia, or the Horn of Plenty.

River myths tend to follow the same pattern. There must be some form of sacrifice that is offered to the river, in this case a bull's horn, and once immersed in the water it will make the river fecund. Particularly, the sacrificial item must reach the bottom, which is symbolic with death, before it can be reborn. This bloody cycle is almost hardwired into early civilisation, as it appears fundamental to the great empires.

The Akkadian civilisation in Mesopotamia has a version of this, predating the Greeks, but, like any myth, it is extremely hard to be precise when the notion first arose or was told by priests or people. Here the god of the harvest, Tammuz, is martyred and the result is fertility. The ritual honouring this cycle involved the slaughter of a ram at the Babylonian new year. Blood would be smeared over the temple before the head and torso of the beast was thrown into the Euphrates. When the spring high water came, a wooden effigy of Tammuz was launched into the river on a funerary boat and sunk. It was the immersion in the water, and settling to the silty bed, that gave life to the crops.

The Akkadian myth was based on the idea of death and resurrection. The ancient Greek myth, in Achelous's tale, is reasonably similar, and the Egyptian one, particularly after the Greeks have spun their own version, has a clear link to the old plot. Diodorus's version celebrates Osiris as the bringer of civilisation, writing, laws, agriculture, all that is good in the world. Importantly Osiris ends things that are 'uncivilised', namely cannibalism. Thus Osiris is a symbol of the civilising process, and a progressive figure, if you like, against backward habits[22].

To Plutarch, the Nile is the 'effusion of Osiris', confirming this sense that the waters were literally seminal[23]. It is unsurprising to learn that, as Seneca writes, drinking Nile water was thought to make barren women fertile. It can be argued that Greece fertilised its own sense of civilisation from the seed of Alexandria. The Egyptian myth and association made Greek culture legitimate – it gave the river of wisdom and power that flowed from Athens a source, an ancient beginning.

Romans were united with Egypt through the bonds of power and empire. The might of Rome stretched to Cairo and Alexandria in mutual support, as the glories of Egypt were envied and acquired by the emperors to the north. Augustus, Hadrian and Caligula all cherry-picked from the monuments of the Nile to decorate Rome.

This wasn't the sympathetic curiosity of Athens, but raw power at work. The emperors collected the obelisks from temples at Thebes and Heliopolis and shipped them back to Italy. We shouldn't lose sight of quite what a task that was – moving a huge stone obelisk across land and sea and then re-erecting it was a sign of ingenuity. It was close to triumphal theft – a calculated vandalism designed to show who was in control. So the obelisks, symbols of the sun and originally raised in worship to the Egyptian sun god Amun-Ra, found themselves trophies in a civilisation with a different set of gods and a different set of values, but driven by the same authority granted by the superior use of water.

The Roman thievery wasn't purely material. They took the idea of the Nile too. Influenced by the wonders of Egypt, and Greece's interest in the great river, the Romans wished to adorn themselves with the same garlands of culture and authority. Temples to Osiris and Isis were common in the Roman Empire. One not far from Rome, built in the first century BC, had a mosaic floor. Depicted in the tesserae is a scene of a great river, rising from within mountains (a challenge there to Herodotus's history) and gushing down over a wide plain, past temples and obelisks and crocodiles and fields. The water's destination is a temple at which a ritual is taking place. A candle is being dipped into a fountain – the Egyptian celebration of the birth of life, the beginning of everything, the sun impregnating the water. This mosaic would have been covered with a shallow slick of water, to make its images all the more river-like and elusive – a presentational trick which not only literally acknowledged the power of water, but which used its shimmering surface to suggest at the mythic, unknowable quality of creation[24].

For Rome, the very idea of great water monuments is a challenge to Egypt, a way of beating them at the business of power and magnificence. The water bureaucrat and historian Frontinus states: 'With such an array of indispensable structures carrying so many aqueducts, compare, if you please, the pyramids, or the famous but useless works of the Greeks!'[25] So for Rome, the aqueduct wasn't just useful, nor merely a symbol of power over other peoples, but a direct challenge to previous civilisations.

For Frontinus, all civilisations worth the name build big things, but Rome's monuments are the best. They are the symbol of Rome's superiority, of how civilisation and progress have advanced under Roman power. This rhetorical hubris roars through the history of civilisation, of one trying to trump the achievements of another. From Renaissance Popes to 20th century dictators, the notion that controlling water indicates superiority is a potent one.

For Romans, the endless flow of clean water on the huge aqueducts, bubbling over fountains and statues and sculpted facades, was all part of the great wonder of civilisation. This hydrodrama expressed an attitude to water. The design of aqueducts barely changes in the history of the empire, but the degree of luxury to which the water is used develops into the opulence of the baths of Trajan and the cascade at Perge. The lesson of Rome, as apparently understood by future societies, was that the powerful not only controlled water, but lavishly wasted it too, allowing it to burble along in public display for no better reason than because they could.

CHAPTER SIX

God, Paradise and Gardens

THE HOLIEST TRIP for the Hindu faithful is to the River Ganges, where they can immerse themselves in the water and renew their purity of faith. Hundreds of thousands gather each year at Varanasi to wash away their sins in the river they believe has flowed straight out of heaven. They hope to be able to stand, semi-nude, in the river and splash its sludge-dark waters over themselves. Even the dead are dipped into the current, cleansed for the afterlife.

Western tourists look on, taking photographs and writing in their travel journals about the spectacle. They may fancy they witness something more 'true' or essential than at home, an ancient ritual that links the people to nature. Part of the appeal to those in the well-scrubbed developed world is that it looks bizarre; the river is dirty, so why would the devout wash in water that cannot make them clean? It looks like a remnant from a pre-civilised world.

Who, after all, would wander down to the banks of the Thames that snakes through London, or the Hudson that separates Manhattan from New Jersey, and plunge in for spiritual purity? Such an act would be odd in societies that value clean water. For periods of recent history, it would be life threatening, given the pollution. In industrialised societies, rivers come to represent commerce, not purity, and the filth that goes with trade. If it is purity and cleanliness you are after in the advanced world, then what's wrong with a power shower?

But the west and the east share a common ancestry here. In many societies, water and spiritual purity are linked, and have been since the beginning of religions. The faithful of rich New York may not dip into the river, but they are familiar with the ritual of baptism. In St Patrick's Cathedral, an imposing Catholic church on Fifth Avenue, one of the wealthiest streets on the planet, the devout will gather to splash water on a baby's head, and so declare the child free from the grip of Satan. Is this any more 'civilised' than the Indian man, loin-cloth dripping from the waters of the Ganges, who believes himself closer to God?

Religion is, despite modern doubts about God and organised worship, essential to civilisation. It tells the story of mankind's journey from the wild to the city and the need for social organisation. Water is essential to many theologies. It plays two crucial roles: First as a sign of how God or gods have imposed order on the world, and secondly as a means for man to start again, to wash away his mistakes, and have another go at being good. If the control of water is at the heart of civilisation, then religion is man's earliest draft of that history.

The roots of Christianity are particularly important. This religion becomes the dominant one in the industrialised world, and as such is most influential in shaping modern civilisation. Christianity's beginnings are almost anti-civilisation. The early believers see the city as the place of sin, and think a wild natural soul more godly. However, as the religion acquires power, it opts in to the same flow as the pagan past. Amidst the pomp of Rome, it learns to value the business of human order above nature, and so begins to forget about water control. This may contribute to the way mankind has travelled on, steadily losing sight of the water pipes as we raise our eyes to wealth and God.

It may now be desert, but the area between the Tigris and the Euphrates was once a swampy marsh. Eridu, the first city, was built on the edges of that damp terrain. It was constructed in a shallow basin, which would flood at times of spring high water. This lagoon was given the name Abzu, which later became Apsu in the Akkadian language.

Abzu or Apsu is the essence of life in Sumerian theology. It also represents the universe. That is, water is the beginning, the thing that comes before everything else. Out of this water comes the earth, literally rising out of the liquid. This emergent solid ground is fertilised by the water, nourishing crops for human consumption.

The idea of a primeval sludge out of which life emerges is a powerful one: we, in our secular age, still refer to the 'primordial soup' as the generalised sense of what gave rise to DNA, the building blocks of life, and all living species. In mythological terms, Absu or Apsu is also a god. He is fresh water. At the beginning he mixes with salty water, the goddess Tiamat. The early version of this is that Apsu and Tiamat give birth to Mummu, a female character who produces heaven and earth, and then the gods of Sumer.

Another version has the water as purely female, called Nammu, and she requires no male to help with the work of creation. However this proves less persuasive on the imagination of Sumerians, who, over time, prefer the idea of a male creator.

We can see how the male god idea is absorbed into culture by looking at the language. 'A' in Sumerian means semen. As noted, it also represents water. The grandson of Apsu is Enki, who is praised in Sumerian hymns for filling river beds with his 'water' – irrigation is associated with, or is the same as, sex and impregnation. Life and civilisation are the bounty of water, and water is akin to an ejaculation from the gods of nutritious liquid. Civilisation is 'born' from water.

Enki is not only the god of water and vegetation, but also of wisdom. So his work produces life and knowledge. The idea of water and semen being equivalent stuff, and of a male creator, will endure and, ultimately win out. Some sources suggest Enki was also the one to gather the waters of the Tigris and the Euphrates, and who governed city life while ensuring agriculture worked.

Water is at the heart of the practice of early religion. It is the most sacred of substances, unsurprisingly in a society that only existed on the control of water. In the middle of Eridu was a temple or shrine. At the heart of that building was a pool of water. We might like to think of it in terms of a Christian font, where babies are baptised. This carried the water for ceremonies. Exactly what rituals were performed is difficult to say, but the indications are that water was the medium for divine action – it carried the prayer or spell, it purified. And, given Enki's quality as an ultimately wise god, the water itself was, literally, the font of wisdom.

Water is also the barrier between life and death. The dead were thought to complete a river journey to the afterlife. The Apsu itself represented the underworld. If you like, the living were above water and the dead below. The Sumerians were quite literal about this – the lagoon around the city was used as a cemetery.

So the earliest known religious practices show that creation was a process of separating earth from the water. Thus the creation myth mirrors a sense of the control imposed on nature by irrigation. The 'civilised' process of putting natural forces, and water in particular, at man's behest is reflected in the myth – the gods must first control water before creating man.

Then we see that the most powerful god is the one associated with water. He has the ability to fertilise the land much as he might impregnate a woman. This deity is also the source of wisdom. All this reflects the practical sense that without control of water, the emerging cities would not survive, as the food that feeds the people would not be grown. So the greatest knowledge – a genuinely life-changing skill – was the knowledge of how to control fresh water.

We should mention in passing horoscopes. The system of relating the movement of the stars to the behaviour of humans was developed in Mesopotamia. It measures history in terms of major 'ages'. Curiously, we have just entered the Age of Aquarius, also known as the water carrier.

The development of religious stories in Mesopotamian civilisations sees a similar cast of characters being given slightly different roles. At the same time another great civilisation was developing along the Nile, and polishing up its religious myth into a tale worthy of so great a power. The dominant narrative is that of Osiris and Isis. This is developed around 2500 BC at the holy centre called Heliopolis, near the city of Memphis, close to modern day Cairo. What we know of this thought comes from the Pyramid Texts, inscribed onto sarcophagi at the pyramids at Saqqara. This is the first known mention of Osiris, and the texts are thought to be the earliest surviving religious writings.

The texts say first there was Atum, a male god. Scholars suggest he was the son of a great female presence, sometimes called Nun, who represents the primeval waters. However, much as in Mesopotamia, the idea of a male life force wins out over the female, and Atum is soon credited as the original deity.

Atum creates out of nothing, or from masturbating, a brother and sister, Shu and Tefnut. Shu is representative, or governs, air and life, while Tefnut symbolises moisture and order. Already, the connection between water and control is made. Shu and Tefnut incestuously mate (thus legitimising the practice of incest for the Pharaohs) and produce Geb and Nut, who represent earth and sky respectively.

Geb and Nut produce other gods, most significantly Osiris and Isis. They also become parents to all the people of Egypt. A powerful symbol in this mythology is the image of Geb (earth), with an erection, trying to get at Nut (sky) and being kept apart by their father Shu (air).

Much like Mesopotamian religious practice, Egyptian civilisation valued the high ground above the water. Heliopolis itself was described as a primal mound. This either is, or symbolises, the centre of creation. It explains in part the cultural fascination with pyramids. These were things which rose up, out of the fecund water, and pointed to the sun, an echo of Eridu's 'holy mountain'[26].

The rituals and beliefs of early Christianity are thought to have stemmed from the Essene cult near Qumran. This is part of a movement that was arguably anti-civilisation. As reflected by the characters in the early books of the Bible, the plain speaking, rural man was always preferred to the city figure, who stood for sin and Satan. These tough, nomadic figures reflect the kind of life led in Qumran. Aridity, scarcity and unpredictability are strong themes of early Christian tradition. The chosen people left the fecund Nile and went towards the Jordan. They adopted the washing routines of the Essene cult, and seem to have developed a sense that free-running clear water is the gift of their god, perhaps in contrast to the tamed channels of other faith systems. If the river is untamed, then the excesses of Alexandria, or Babylon, are avoided. God didn't like the city, or the Nile[27].

While the Nile was a symbol of strength for others, and the red waters celebrated by Egyptian priests as a sign of plenty, the early Christians used the same image for other purposes. In the Hebraic Book of Exodus, the flight to the promised land out of Egypt is because the river has turned to blood. The colour of the Nile was taken to be a sign of God's wrath, the famines and plagues suggestive of the Nile's low periods, proof that the one true God didn't approve of this fluvial culture. The early Christian church used the Nile as a symbol of all that was wicked: the calm Jordan is seen as a godly alternative.

The great threat to civilisation was flood. The triumph of irrigation was not only that it gave man reliable harvests, but it stopped the floods washing them away. That is why flooding occurs in the earliest religious myths. The clay tablets that record Sumerian creation tell of a great flood, sent by the gods, who had become exhausted by man's failings and wished to start again. Only one man and woman survived this first watery destruction, Ziusudra and his wife, and they did so by building a boat, or what could be an ark.

This crops up in similar form in the Babylonian myths, along with the idea of the ark carrying many living things. In the 'Epic of Gilgamesh', Ea the god of wisdom (and water), tells the king of the city of Shurrapak that the gods are determined to destroy the fallible, irritating ways of man.

> Man of Shurrapak, he said, tear down your house
> And build a ship. Abandon your possessions
> And the works that you find beautiful and crave
> And save your life instead. Into the ship
> Bring the seed of all the living creatures.

The gods do this because they do not wish to give man access to the city any more. That symbol of the bounty of irrigation, the city, the wonder of civilisation, is the thing which man is denied. And the chaos of flooding follows:

> Then the gods of the abyss rose up
> Nergal pulled out the dams of the nether waters,
> Ninurta the warlord threw down the dykes...[28]

The flood does two crucial things: it cleanses the earth and humans of the wrong that has occurred, and gives them a chance to begin again. The flood starts out as disaster and becomes salvation. The 'Epic of Gilgamesh' also gives man some useful advice – don't build your home on a flood plain, but at the source of the river. Floods occur downstream.

In natural terms, a flood is a new beginning. That it clears the earth is literally true – any flood will destroy the things in its way. It will also leave a bed of silt, the topsoil from upstream that has been washed down. This is rich ground, and is good for planting in. It is for this reason that the Egyptian myths do not hold floods to be disaster, and have no tale of man escaping the deluge. When the Nile floods, it is good, as it provides the nutrient heavy earth in which crops will grow. That is why Hapi the river god is celebrated for his floods, and 17 June is marked by a celebration for the 'night of the drop', when the waters are said to start rising.

But the reality of irrigation sets in train an idea, that floods can be controlled. In the early religions, it is God who is in charge of the waters. A few thousand years later, with the age of reason, and man's steady process of replacing God with himself at the centre of things, it will become an obsessive idea that man can hold back the floods. This will prove to be a powerful delusion.

The Hebrew religion recycles the flood story from earlier myths. This time it is Noah who is told to build an ark, and fill it with other creatures, which he does, and when the waters come they drown everything, and ultimately he lands again on Mount Ararat (another mound coming out of the water).

So the idea of water as the purifier has taken root in religion. There is a direct link from Noah's flood cleansing the world to the baptism service in St Patrick's Cathedral in modern day New York. The water splashed three times on the head of the baby will have washed away Satan's presence, a quiet echo through time of the Flood drowning out the mistakes of man.

For early Christians the city is not necessary, but the garden is – the heroes of the early church forswore the luxuries of urban wealth and sought sanctuary and enlightenment in the wild, always trying to get closer to the idea of paradise, before the fall, to the original garden. Yet the Garden of Eden wasn't some new religious idea, but integral to the history of civilisation, and man's control over water.

The garden is a product of water control. The powerful replicated their mastery over the fields through irrigation with smaller plots, where the water was as integral to the experience as any plant. Cyrus was the founder of the Achaemenid Empire and he built a palace at Pasargadae in the sixth century BC, the remains of which are the earliest surviving evidence of garden design on record. There is a water border, and the layout is in four sections, split by water channels. This is the first extant example of what is a classic garden form – the water controls the space, marks the edges, and is as much a part of the aesthetic as the plants. Later this geometric pattern will assume religious significance, but it is a product of irrigation. The channels need to be at right angles for the masonry walls not to leak, and the water is there in the first place for flood irrigation, as there were no hoses or sprinklers.

Obviously such a garden would appear luxurious. The water borders were a clear sign of the wealth and power of the owner. Perhaps equally obviously, they were associated with rest and reflection. In a world of hard labour, short lives and basic survival, to have a refuge of fruit trees, quiet water and tranquillity couldn't help but suggest something more spiritual. It is little wonder early religions would adopt them as special places.

The Persians referred to enclosures, which ranged from small gardens to hunting parks, as 'Pairidaeza'. This word is a compound, literally, as it derives from 'pairi' meaning around and 'daeza' meaning wall. The Greek writer Xenophon translates this as 'paradeisos', which is in turn used in the Greek translation of the Bible, when describing the Garden of Eden. It is from this that English gets 'paradise'.

The word is a shadow of the fact, and enclosed gardens are understood to have occurred in the earliest of the Mesopotamian cities. Archaeologists believe you may find a rose and a fountain near the centre of Ur, just as you might in a modern suburban yard. The fountain showed man had water as his servant, the fruit and flowers proof that this was a happy relationship.

In fact early representations and descriptions of paradise include cities and island trading posts but it is the garden that wins out. This may reflect a simple economic truth, that the rich had gardens and the poor didn't. They were literally cool places, where men went to think higher thoughts while listening to the gentle flow of water.

Xenophon toured Assyria and Persia around 400 BC, reporting back on the verdant luxury to the east. He writes in 'The Oeconomicus' of one king, probably a descendant of Cyrus the Great, that '... in all the districts he resides in and visits... takes care that there are "paradises" as they call them, full of all the good and beautiful things that the soil will produce.'[29]

Christianity and Islam build on the long tradition of garden paradises. In so doing, they put the idea of controlled, managed nature at the heart of civilisation. The flower beds, lawns and fountains may symbolize pure wilderness, but one tailored to our aesthetic and obedient to our whim. Coupled to the effort to hold back floods, the idea that nature can be improved upon will have major consequences for how man relates to the environment. Paradoxically, the search for paradise is the quest to dominate nature, and at the heart of it lies our relationship with water.

For example, to sustain the gardens of what is now Iran, special channels were developed to get mountain water to the dry plain. No doubt this water would have been used for other purposes too. These channels are called qanats, which are underground tunnels cut lower than the water table. They tap aquifers so that water pours for miles towards gardens and cities. As it is covered, the qanat prevents the summer sun evaporating most of the flow. The tunnel system delivers such an important commodity that the bureaucrat in charge of its maintenance is called a 'mirab' – a 'prince of water'. It is an effective system. Without it the great desert city of Persepolis would have been unable to bloom. The technology is adopted in other regions and by other empires, eventually becoming one of the first ways of bringing water to the distant Spanish colony in the New World – Los Angeles[30].

How the Spanish in America ended up with this Persian technology is a tale of two religions. The civilised world of 1,500 years ago was a small place. It stretched around the Mediterranean, through Persia to the Indus. China and Mesoamerica were not considered part of the club. Trade and towns lined the Arabian gulf, but inland nothing was thought worth having.

Only recently has the vast desert of Saudi Arabia shaken its colonial name of the Empty Quarter. With searing heat and no water, it was naturally no place for civilisation. Around 500 AD it was home to nomadic tribes. This meant there was no organised government or social order. The only towns as such were in Mecca and Medina on the western coast. In Mecca, there was a shrine of black stone, surrounded by a holy area. A tribe called the Qarysh ruled, of whom Mohammed was a member.

Around 600 AD Mohammed said he was visited by an apparition who told him there was only one God. Mohammed was told to preach this, and so he did, echoing the monotheism of the Abrahamic faith. This presented quite a challenge to Mohammed's neighbours. They were used to pledging allegiance to their tribe. It is typical of nomadic peoples that community is defined as bonds of blood, not an abstract social contract. Mohammed challenged this by suggesting the family of the faithful was superior. Islam imposed a social order on nomadic people.

Part of the incentive for this was the notion that to be unfaithful was to invite hell in the afterlife. No other religion has quite such a fixed and definite notion of what happens after death. The Muslim faith was specific on the notion that a paradise awaited the godly. Translators are hazy as to whether this paradise was populated by virgins or good wives, but what is certain is that it was a garden. The paradise mosaics at Umayyad Mosque in Damascus depict this heaven – though interestingly they also show city walls, a reminder that early ideas of paradise included the sense that the urban was a good thing.

This fixation with a mythical garden echoed the genuine spoils of war: as the Muslim armies went north into Persia, Byzantium and Assyria, they went into lands rich with fruits, nuts, wine and grain. This is no happy side benefit. The garden and irrigation became integral to Islam. The faith was built around the promise and reward of making the desert bloom.

This is why water is often depicted in Islam as green. That is the colour of what water does – it makes things grow. Green is the garden, the oasis. Further, Islam shuns representational art in favour of patterns. Among these are the patterns of gardens. The forms of Pasargadae are echoed throughout the millennia in the design of carpets, which are directly derived from the layout of the classic model of paradise. Religious teaching layers theology on top of the water – the classic garden design with four channels comes to represent the four rivers of life.

Further, Islam takes the garden as the crowning symbol of its superiority. As the Islamic armies spread, first north and then east to the Indus, and ultimately west along north Africa and into Spain, they make the garden the focus of their palaces. Paradise comes with them.

A very different thing is happening in Christianity. Paradise is an abstract concept for the theologians. It is a vague promise. The Garden of Eden may represent what life was like before the Fall, but there is no clear promise that heaven will be thick with fruit trees and fountains. One influence Christianity absorbs which Islam does not is the wet north. Though the garden remains a luxurious retreat and an object of obsession by the rich in Europe, its promise of fruit and greenery is much less potent in a land where food is in abundance. So paradise is internalised.

Christianity's dominance in England, France and Germany is expressed through the spread of monasteries. These develop herb gardens, and the Cistercians in particular are adept at constructing watercourses through monasteries, bringing in drinking water and taking waste away. But the decorative or pleasure garden also occurs, and adopts the same pattern as Pasargadae, or the Islamic layout. There are water borders, four segments divided by water channels, and fountains. In a wet country, the fountain should be redundant, comically pointless, but it succeeds.

In 1260 Dominican Albertus Magnus wrote on what a pleasure garden should be – thought to be copied from Bartholomew de Glanville's encyclopaedia written 20 years earlier – 'Best of all were the gardens with a fountain in the middle, tinkling into a stone basin.'[32]

None of this is paradise though. For Christians, that is something indoors. The church or cathedral is the paradise on earth, the soaring stonework an expression of God's glory, not the blooming flower. It is in the intersection of rib vaults and the decoration of the façades that the Middle Ages grows their heavenly roses – stone flowers which will be in bloom for eternity.

The ground plan of the cathedral at St Gall in Switzerland, built in the ninth century AD, shows an area called 'paradise'. This is the colonnaded area around the apse. It is tempting to think of the word 'apse' as being derived from 'apsu', but it appears to come from the Greek 'apsis' meaning 'fitted together'. These columns might have been borrowed from Sicilian church design, where sheltered areas were used for the growing of flowers for religious festivals. The ancient paradeiso (literally a 'wall around') has become a wall around the end of the church[33].

Similarly the iconography of Christianity takes a divergent view from Islam. There is not the strong cultural link between the effect of precious water turning the desert green. Instead water, crucial in religious ceremony as the purifying agent of God's will, is valued for its purity. Nothing in Christianity is more pure than the Virgin Mary, who by immaculate conception carried and gave birth to Christ. The emerging palette of representational art in early Christianity gives images of Mary in a blue gown. Blue is pure. Water to the north, despite in reality being any number of murky hues, is portrayed as blue.

Both traditions merge in Spain. The Moors were in the Iberian Peninsula from the sixth century AD on. There, they construct some of the finest Islamic palaces and places of worship. For Islam, paradise is still the enclosed garden, the wall that keeps nature safe from the uncertainty of the arid world beyond.

The cities of Cordoba, Seville and Grenada are the lasting legacy of Moorish water ingenuity. Water and civilisation are so obviously the same to a Moor that the two blend without question. The irrigation system serving the mosque at Cordoba is modelled on the interior of the building. The palace at Generalife is integral to the gardens – to separate interior from exterior is aesthetically impossible.

The Moors developed sophisticated irrigation and water collection systems. In the Middle Ages, Cordoba, capital of Moorish Spain, was able to support a population of one million people through irrigation. At the same time, the largest city in northern Europe was London, with 35,000. However, superior hydro-engineering could not repel the Christian armies of the north, which gained control in 1492. They drove the Moors out of Spain and back to north Africa, the gardens and irrigation systems left behind as a permanent reminder of the glories of Islam. However, Queen Isabella II was little impressed by these achievements, and commanded that the water wheel of Cordoba, which had kept the city's irrigation functioning, be taken down, as it kept her awake at night.

It was from Christian Spain, a place that took for granted the connection between water and civilisation, that Columbus set sail in search of a quick route to India. Instead, he found America. Had it been a Moor who had made that journey, then world history might be very different. As it was, the prize in mind was trade and personal wealth – paradise through money. Water had yet again sunk from view. Our conceit of what made civilisation was moving further from the simple business of controlling the precious liquid, and becoming more entangled with notions of human superiority and power. The relentless thirst of civilisation would increase, but our minds would be elsewhere, and the trouble stored up for another day.

II

THE GROWING THIRST

THE SEABED IS BEING dug up and dumped in dry piles off the coast of Dubai. These sandcastles will become housing, sold to the wealthy as a luxury retreat and a good financial bet. The scale and ingenuity of the means is impressive, even if the end is slightly baffling to my eye – buying a house built on sand seems like an investment destined to fail. A 14th century Dutchman would share the sense of amazement, and perhaps claim that there is nothing new in the world. Hundreds of years ago the Low Countries were reclaimed from the sea, and in the process the modern state was begun.

This section of the book is about the growth of civilisation, or how the western state and way of life emerged from the waters to become the major force in the modern world. We can only estimate, but at this point in history, man was drawing water at a very slow rate. For every bucket pulled up the well, nature rained down far more water to replace it. The aqueducts of Rome, or the fountains of Moorish gardens, were a tiny part of civilisation's small scratch on the earth. The next thousand years would change this beyond measure.

Imagine a glass raised to the lips of mankind; at first we only sipped the water. From the emergence of the Dutch Empire on, we began to swallow mouthfuls. This section is about how we learned to gulp down our most precious resource.

CHAPTER SEVEN

God's Dry Land

AS CHILDREN MANY OF US were told the tale of a brave boy who saved a village from flooding by sticking his finger in a dyke, preventing the water from bursting through. Like other childhood memories, it comes back to us incomplete; How long did this poor lad stand there, his digit wet? My mental picture had the boy diving in and blocking the leak from the waterside of the dyke, leading me to think the story was less about heroics than the drowning of a small child.

This primary school tale is one of the shreds that remain in popular culture of a civilisation that changed the shape of the world, and set the tone of modern, liberal capitalism. This was the Dutch empire, the first of the truly great west European power blocks, which created a society of previously unimagined wealth, fairness and liberty. It built this achievement, and its sense of national identity, by controlling water[34].

Travellers to the Low Countries from the late 16th century on wrote about a strange Dutch phenomenon, a way of punishing criminals that was so barbaric it couldn't fail to shock the reader. This was the drowning cell. According to the horrified, but fascinated, prose of the alleged witnesses to this device, it consisted of a cellar, a wooden post and a water pump. The miscreant was tethered to the post in the cellar, then sluices would open and the small space filled with the water. The only chance of surviving was to pump as energetically as possible, in order to prevent drowning.

No travellers' tale was complete without the gory details in the 16th and 17th centuries. This still has the resonance of a grotesque torture. Unfortunately, the historical evidence for the drowning cell's existence is hazy. Otherwise known as the 'water house', it was allegedly sited in Amsterdam. Records show no such punishment ever being carried out.

That is not to diminish the importance of the story; it gripped the imagination because it fitted with popular ideas about the Dutch. Two great revolutions were occurring in the Low Countries, both of which suggested the people were tougher and smarter than others. One was the battle against the sea – no civilisation had ever waged such a concerted campaign against the force of nature before. The other was religious –

this emergent nation and empire was doggedly rejecting the single faith from Rome, and sticking to its Protestant guns. The issue at the heart of the nation's success, water control, seemed also a metaphor for a new kind of human resilience and triumph.

To drown was to fail, to be judged, in 16th century Holland, while to escape the water and beat its lethal flow was to be saved, not just in the eyes of society, but before God. As the historian Simon Schama puts it: 'To be wet was to be a victim, to be dry was to be free.'

The heraldic motto of Zeeland, the largest region of the Low Countries, was 'Luctor et Emergo' which means 'I fight to emerge'. One might like to imagine this as emerging from medieval Europe, or from a thousand years of gloom when the bright light of Roman civilisation dimmed in the damp, northern bogs. The Dutch may also have fancied it meant emerging from a corrupt spirituality to the dry certainty of Protestantism. But the phrase is literal; their land existed because of a continual struggle against the sea. They had beaten the flood, in an ark whose walls were the earthen dykes which held back the tides, and they had been delivered to the promised land. They were dry, and so they were the chosen people.

The Low Countries are accurately named. Much of the land on the western fringe is at, or below, sea level. High tides or rough storms would drive the salty waves over the land, killing people, flooding villages and ruining the fertility of the soil. If the folk of these boggy, exposed territories were to survive, they had to hold back the sea. So they became the world's experts at building dykes.

A dyke is a raised ridge of earth. It is sloped on both sides, like a small hill, and runs the length of either a riverbank or a coastal edge, holding back high tides or storms. It is arguably the simplest form of water technology known to man – like creating a ridge of sand on a beach to protect your bucket-shaped castle; unfortunately the real thing can be just as fragile.

On St Elizabeth's Day in 1421 a ferocious storm whipped the North Sea into white-capped peaks of rage. The dykes constructed along the Zuider Zee broke, and the water rushed in, killing perhaps 10,000 people. By morning on the following day, that which had been fields and villages was sea again, for an area of 500 km^2. This tragedy imprinted itself in the Dutch psyche – this was symbolic of all the floods that made life so hard. The St Elizabeth's Day inundation represented the water that had broken into people's lives throughout the Middle Ages and before: a folk memory of the sea beating back people's struggle for life and drowning their dreams.

The Dutch knew the vagaries of the sea's domain better than most. Over time the simple earth dykes that lined the Zuider Zee would be replaced with more sophisticated technology. Amsterdam was growing fast in the early 1600s, as cities do when the food supply becomes regular, wealth appears to be stable and population rises. So a syndicate was formed to drain the Beemster area. The boggy land was reclaimed: 43 windmills powering water wheels lifted buckets of water into canals, creating 70 km^2 of new land. This would help provide the food for Amsterdam, which grew from a population of 30,000 in 1580 to 150,000 by 1650.

As in Mesopotamia, Egypt or the banks of the Yellow River, a civilisation arose by controlling the flow of water. Much like these earlier settlements, the impetus for civilisation came from struggle. Where there is naturally plenty of food, and the water source is reliable, and predators are kept within reasonable check, there is no motive to civilise. After all, if one is well fed and comfortable, why bother collaborate with your neighbour on huge, laborious endeavours? The struggle to push back the sea forged the Dutch nation. Where there had been salty pools, there were now cut ditches and canals of fresh water. The first large scale, man-made farmland in Europe was created.

The Polder Het Grootslag is a painting by an unknown artist dated around 1595 and it is in the Rijksmuseum Zuiderzee Museum in Enkhuizen. It shows regular plots of rectangular fields. Some of the land is planted with crops while on other areas cows graze. Small boats ply the narrow drainage canals while confident, well-to-do men stand in dark clothes and white ruffle collars surveying their enterprise. This uniform landscape stretches to the horizon. Remove the water and the windmills, and you have a template for the mechanised farmlands of today, to be found across the developed world.

And from this ordered land sprung a nation of previously unimaginable wealth. The Dutch had not only converted their sea-washed soil into a rich ground that fed the citizens with plenty, but they looked far beyond their own shores for opportunity. In establishing the spice trade with the Far East, whereby ships from the Hague sailed to modern Malaysia and Borneo and came back laden with pepper and cloves, the civilisation of the Low Countries also set the template for modern capitalism. This wasn't acquisition by military power and low cunning, as rival European nations were adept at, nor wealth from brute strength, as the Viking raiders had used. The trade in spices from the tropics of Asia to the cold north west of Europe was a step-change in human relations. Where previous civilisations had traded in their known world – largely the shores of the Mediterranean and the lands of Mesopotamia and the Levant – the Dutch initiated commercial links with the other side of the world, and did so on the promise of mutual benefit. The Dutch developed a society that would be the foundation of modern living.

The English traveller William Anglionby was one of many who came to the Netherlands to marvel at the perfect state born amongst the canals and windmills. In 1660 he found himself in Leiderdorp, a village on the outskirts of Leiden, where his mouth was agape with wonder that a small settlement like this should have 'more palaces than country people's houses. Tis here where we must admire the magnificence of the citizens, for one would think that there were an emulation between who would show most marks of riches by their expenses.'[36]

These were the robust finances of the world's richest state. The Hague, according to Anglionby, 'by the breadth of its streets, the nobleness of its buildings, the pleasant shade of its trees and the civility of its inhabitants, may justly claim the title of the most pleasant place in the world and may make all men envy the happiness of those who live in it.'

What could be more civilised than the notion of an ordered, wealthy city in which the inhabitants were actually happy? Most European cities were squalid mazes of disease, crime and decay. Yet here, in this ostensibly damp corner of the continent, was a new, urban paradise. Anglionby said of Amsterdam 'this city is very like Venice. For my part I believe Amsterdam to be much superior in riches.' Exactly 100 years later another traveller from England, Joseph Marshall, summed up his experience of Dutch society thus: 'In a word you view not only the conveniences of life, but those improvements, those refinements which rich and luxurious ages only know.'

The Dutch credited God and hard work for the bounty. This fitted in with the emerging religion of Calvinism. It not only values honest labour, but also subscribes to the idea of predestination. This is the theological idea that some people are chosen by God and they will go to heaven, while others are not and no amount of praying will change their fate.

It draws on the Old Testament view that the Israelites are the chosen people. Noah is obviously one of the elect, as Calvinism would describe it, while those who drowned in the flood deserved what they got.

Though well under half the population of 17th century Holland were Calvinists, the creed became adopted as a part of the national identity to distinguish the Dutch from their enemy, Catholic Spain. That explains why, in one of the first attempts at a national anthem, the 1626 Neder-Lantsche Gedenck-Clanck (The Netherlands Anthem of Commemoration), describes the people as water-savvy:

> O Lord when all was ill with us You brought us up into a land wherein we were enriched through trade and commerce and have dealt kindly with us, even as you have led the Children of Israel from their Babylonian prison; the waters receded before us and you brought us dry-footed even as the people of yore, with Moses and with Joshua, were brought to their promised land.

Allowing for the hubris in all national propaganda, here we have a clear expression of the link between water technology and the people's sense of entitlement. They deserved their wealth and dry land, because God had decreed it. Therefore it was an easy step to thinking that controlling water was the work of God. The 16th century water engineer Andries Vierlingh offers this humble assessment of his own talent – 'The making of new land belongs to God alone, for He gives to some people the wit and strength to do it.'

This makes it sound as if the Dutch not only adopted water technology and so civilisation, but also subscribed to the quasi-religious view of society, replacing the gods of Sumerian, Egyptian or Roman culture with the singular, Protestant one. This may suggest an autocratic society, where the individual was subservient not only to the greater will, but the will of God at that. In fact nothing could be further from the truth. The Dutch created a culture on the back of their mastery of water which not only laid the foundation for modern enterprise, but also for the rights of the citizen. For the first time in history we see the idea of the individual emerge, a person whose fate, despite what predestination may say, was actually bound up by their relationship to Mammon, not God. The roots of this lie in water, too.

Across Europe the social structure was feudal. In simple terms, this means that the monarch derived personal authority from God. In turn, the King or Queen would empower lords to govern regions of the kingdom. These lords became owners of vast tracts of land, which would be worked by people, who would establish their right to plough a certain field or manage a wood by paying the lord a 'feu'. In practice, the notion of ownership was very loose indeed. So long as the lord was paid, what did he care if he 'owned' this field or that; he had no way of using the land other than to let the farmer get on with farming. There was also a lot of common land, which everyone owned. This was often boggy or wooded, and had limited economic potential for the lord.

Three factors forced this system to change: forest clearing, land drainage and the enclosure system of farming. If you enclose a flock of sheep behind a wall, as opposed to letting them run wild, it's easier to look after them. That means you have more sheep and more to wool to sell. Which means the value of the land enclosed has risen.

Now that land can have an improved value, there is an incentive for the lord to become more active in 'owning' it; why let a peasant pay you some measly rent when you can put sheep on that same patch of soil and earn more. So the abstract notion of 'ownership' suddenly became very real indeed. Lords were driven to cut down forests and drain land in the pursuit of acquiring more money. Once they invested cash in the effort of wood chopping and ditch digging, they assumed ownership of the land.

Across forested and damp Europe, people were discovering that they no longer had access to their commons, and were not welcome on their homeland, as the wealthy were busy getting wealthier by developing the notion of land ownership. When the traditional feudal obligations were abolished in 1549, the Netherlands had an alternative model of land ownership ready.

Cutting down forests is a relatively cheap thing to do. You need strong men, some axes and saws and horses to pull away the timber. Fire will burn away the brush and roots, leaving you with ground that can be cultivated. Draining land is very expensive. You need a lot of people, with a lot of time and energy, and equipment to dig ditches, move soil, build lock gates and bridges. Ancient civilisations were able to organise armies of labour through the promise of food. In regions where food was scarce, people were willing to collaborate on irrigation schemes in the hope of regular meals.

This does not carry the same potency in northern Europe, where food was in plentiful supply, growing and running wild in the common land. That is also why the peasant did not feel so indebted to the lord – unlike in Mesopotamia, where the powerful controlled the water, and so controlled the flow of food, in Europe the peasant only relied on the lord for military protection. It was a less desperate relationship in the first place.

Therefore a Dutch lord would have had limited success simply ordering the people on his land to drain it – the people might have been wet and dirty, but they probably were not hungry. He would need to pay them if the work was to be done. Given that work was very expensive, individual lords were unlikely to have the cash to afford such grand schemes. What emerges out of the sea is a new model of ownership. The feudal model gives way to syndicates of investors. So the Beemster project for example would have been funded by a group of wealthy Amsterdam citizens. Their money bought the labour and the material for the windmills and canals. In order to see a return on their investment, the syndicate charged rent to the occupants of the land. Overseeing all this was a 'Water Guardian', an official who organised the water channels. This leads the Dutch to think in terms of small plots of land with specific cash value, rent, attached. It is a small leap to buying out your plot entirely. The smallholder was born on the flatlands of Holland.

This eroded the feudal system, replacing it with a commercial structure. Your right to a piece of land was determined not by a sense of duty to a lord or the Crown, but by cash. This had a profound effect. While the rest of Europe was entering into a 200 year period in which one of the central themes was the tussle between citizen and king, the Dutch evolved an alternate system where the idea of the individual governing their own destiny in society according to how much they earned took hold.

Contemporary observers from abroad would marvel at the Dutch system of government. Though I have referred to things such as the creation of a national identity, there was no literal nation here. No king ever united the Low Countries by battle or through judicious marriage, nor was there some founding moment, as captured in a treaty or declaration. A loose conglomeration of regions and smaller states were bound together by their struggle against the sea, and the resistance to outside aggressors. They had a monarchy, the House of Orange, but no simple structure of government, such as a court and a parliament. Instead there were multiple bodies of authority. This system could only occur, and work, in a place where traditional European social hierarchy had been replaced by a sense of individual merit. Though they had collaborated to fight back against the sea, this process had spawned not the autocratic societies of elsewhere, but an egalitarian one, where money was the mark of personal success. Who needed a feudal lord to protect you when the enemy was nature? Much better appoint a 'Water Guardian' and pay for the service.

The Spanish discovered to their peril that these storm-battered people had a different way of doing things. Charles v tried to create a centralised body for water management. He may have thought he was bringing order to a chaotic system, but his actions provoked anger. The fiercely independent people from the Low Countries resented this attempt at 'foreign' government and the loss of local identity. The tax that the Spanish king applied was also resisted; if it was being collected in a central pot, who could tell if it was being spent on water schemes, as intended, or siphoned off for nefarious purposes. The new water agency ignored the large body of water law that had developed over time in the Netherlands, a jumble of statutes that amounted to a codifying of Dutch identity – independent, local, proud, subservient to no one. Unhappiness over this measure contributed to the revolt against the Spanish rule and the formation of the Dutch Republic.

What occurs in Holland is the creation of something close to our modern understanding of the citizen and the state. While crown and church have their roles to play, the crucial contract is between the individual and the greater good. This evolves out of the water schemes which required the collaboration and good will of everyone, and which give rise to a new power relationship, based on money not feudal obligation.

CHAPTER EIGHT

Fountains of Rome

WHEN ANITA ECKBERG STOOD in the waters of the Trevi fountain, she shivered for her art. Marcelo Mastroantonio had a wet suit on under his sharp tailoring, and gulped at a bottle of vodka to keep him warm, but the blonde beauty spent the night in the fountain's spray with no more than an alluring smile and a little black dress.

The image from the Italian film *La Dolce Vita* is an iconic one, so much so that tourists at the Roman fountain still have to be discouraged from re-enacting the famous scene. The movie revels in contrasting the sharpness of modernity against an historic background, it is a hymn to a very European sense of uniting the ancient past with a glamorous future. Eckberg, the personification of northern beauty, is immersed in a pool that owes its roots to the Renaissance, ancient Rome and ultimately the power of the Nile. This is the sweet life, as the cultural river that flows around her body is as rich as the Nile's sediment, and she is a bounty as great as the wheat and dates that grew on the irrigated lands of Egypt.

If the idea of individualism and personal property was being developed in the north, then the grand accoutrements of civilisation, the symbols and quality of it, were being polished in the European south. Italy went about the business of reviving the ancient civilisations, while blending them with new learning. In so doing, a kind of cultural standard was created to which the global civilisation would always aspire. This asserted, amongst other things, the value of flowing water as a symbol of success.

The Trevi may have played a leading part in the movie, but the fountain which best tells the story of civilisation is a short walk away in the Piazza Navona. Flowing water and knowledge have long been connected. The Greeks made the link explicit by developing the idea of a 'font' of wisdom. The modern world of 17th century Rome builds its own font, in the form of the Fountain of the Four Rivers. Here the ancient fascination with the Nile is bound in with the mythic power of rivers in general, and is presented to the people of Rome as a testament to the civilising glory of the Renaissance.

The Fountain of the Four Rivers was designed by Gian Lorenzo Bernini.

He was an archetypal Renaissance Man; an historian, architect and writer. His plays were satires and comedies, staged in the Palazzo Barberini. One was called *The Flooding of the Tibre* and involved a *coup de théâtre* – water gushed from the back of the stage towards the audience, only to be diverted just as it appeared about to flood the laps of the stunned patrons. One can't help thinking of Bernini's audience like those who witnessed the early films, shocked by the illusion of speed and danger. Water was trickery and entertainment.

Bernini understood this, and offers us a sense of how modern civilisation has 'forgotten' water, lost sight of its central importance. Who among modern architects or artists would think to describe themselves, as Bernini did, as 'un amico dell'acqua'. To be 'a friend of water' would be a strange and meaningless thing to say now, but it signalled, in the 17th century, an understanding of the origins of civilisation.

The Fountain of the Four Rivers manages to combine Egyptian, Roman and Christian traditions all in one. Out of a torment of rock, a struggle of solid matter and muscular gods, rises a perfectly geometric obelisk. The sharp stone points to the sun and beneath it the baroque roll of apparently primal rock is pierced by air and water. It is as if all the creation myths were rolled into one big bang of joy.

Where fountains traditionally spouted from the top, here the water squeezes out through cracks in the lower tier of rock. Out of this elemental fusion spring the four rivers of paradise. Civilisation's acquired knowledge has expanded beyond the reach of the Euphrates, Tigris and the Tiber. Here the mighty rivers are drawn from separate continents. The huge, muscular river gods at each corner represent the Danube, the Ganges and the River Plate, or Rio de la Plata. The fourth god is, of course, that of the Nile.

The obelisk that rises like a shaft of light above their tortuous effort is genuinely Egyptian. It had been brought to Rome, as so many had, by an emperor wishing to illuminate his reign, and ego, with the light of the Nile. Domitian had shipped the plinth of stone a thousand years before, and it had initially stood tall, but for this fountain it was rescued from neglect.

While there are differences of opinion on precise meanings behind each element, there is no doubt the fountain as a whole was meant as a statement. The message was clear: whoever commissioned this sculpture was the master of both the past and present, and was the font from which current power ran. The Renaissance popes instinctively understood the origins of their title. 'Pontifex Maximus' was what the Romans had called the engineer in charge of the first bridge across the Tiber. 'Bridge Builder in Chief' is what these pontiffs were – linking the accumulated wisdom and potency of the past to themselves. By appropriating the symbols and emotional currency of previous civilisations, they hoped to show the divine rightness of their own power. If they were to be, demonstrably, God's representative on earth, they had to show how they were the masters of the pagan gods too. This may appear contradictory, but it's an essential process of civilisation – adapting the past to suit the political needs of the present. And the papacy was a master at such high politics.

Any Christian scholar knew the roots of the one faith lay in many pagan ones, and in the achievements of previous civilisations. The job facing a newly empowered and unchallenged church was to draw those sources together in a single current. This was a contradictory business, showing how Christianity was superior to other faiths and the past, but relying on the imagery and myth of the pre-Christian world. The popes were enlisting the authority of previous civilisations to lend weight to their own claims of power.

A thorough reading of the classical texts wasn't necessary to understand the innate superiority of ancient Rome over its medieval child. All around was evidence of a civilisation that had mastered water to a higher degree than anything achieved by contemporaries, and the city was littered with stone monuments erected by the great powers a thousand or more years before. If a pope, or anyone for that matter, was to show they were the inheritors of the great tradition, then they had to revive the ancient city.

The statue of the river Nile was originally at a temple to Isis in Rome. The figure was rediscovered around 1500, along with another reclining God, this one an effigy of the Tiber, carrying a horn of plenty, the original cornucopia. The two were moved to a prime location by Michelangelo and placed at the base of the stairway of the Campidoglio. The potent river gods of the ancients were now the guardians of the Renaissance[37].

It is one example of the popes adopting Egyptian iconography and associating themselves with the power of the Nile. Julius II, Alexander VI and Sictus V all joined in a process of venerating the monuments of old in pursuit of a new currency of power. Sixtus V re-erected obelisks, brought over by Roman emperors, on Christian sites. He wanted to use the mystery of the Nile to further the claim of God, and by extension, himself.

Nicholas v had also initiated a programme of reviving Roman water technology. Aqueducts were restored. Sixtus v took this further, and imagined a new greening of Rome, a revival of the city through new irrigation allowing for a greater population. Sixtus embarked on a scheme of new pipes, baths and pools. Water was the rejuvenating element for the papacy and the city. Paul v continued the momentum, repairing the great Acqua Trajana. Water that had once flowed to the imperial baths now spouted out over Christian imagery. So the great city had enjoyed a water-driven, Nile-revering renaissance for over a century before Bernini's masterpiece was even conceived.

Bernini had carried out major works for Pope Urban VIII, and was the favoured creator of the time. Urban was from the Barberini family, and when he died his family and friends were run out of town by envious rivals. The Pamphili family benefited from the vacuum of power, allowing Innocent X to be installed as pope.

Innocent wished to leave an indelible mark on the city. This meant more than the building of a church or a fountain, which would presumably outlast him. It meant claiming a bit of the city as his 'turf' so to speak, a quarter of Rome that would always be associated with him. Innocent chose Piazza Navona. He wanted his favoured architect Borromini to expand the pagan temple in the Piazza into Sant'Agnese, and he wanted a great big fountain in the middle. This market space, witness to a throng of traders and citizens every day, would be dominated by a glorious testimony to Innocent X. He would bring the Acqua Vergine into the square and it would gush up amid the people, babbling about his power to all who looked.

In contrast to Borromini's conservative design, Bernini conjured up a theatrical masterpiece which proved irresistible to Innocent's ego. The pope wanted to re-erect an Egyptian obelisk, in a square which had once the been the site of pagan games, and declare the new monument a testament to God, and do so in time for a religious festival in 1650. The obelisk rises out of the twisted rock, which is pierced on four sides, so that air and light pass beneath the needle of stone. This in itself is a great illusion, a 'wow' moment for any onlooker. Here the magic of the Nile is reconciled not just to Rome, but to all of civilisation. The figure of the

Danube shows the new military and diplomatic links with the Church, in Rome, and the Holy Roman Empire, based in Vienna. The Rio de la Plata figure symbolises the New World, and the virgin territory for Christian conversion. The Danube sits between the Rio de la Plata and the Nile, signalling the link between the ancient world and the new one, the bridge between civilisations, and how they run together. The Ganges is symbolic of the fourth continent, Asia – this is the known world in stone and water. On top of the obelisk was a dove carrying an olive branch – like the angel on a Christmas tree, a Christian symbol crowning a pagan tradition.

Renaissance Rome took the traditions of the past and blended them into something new. It dressed the dark and mysterious Nile civilisation in the new fabric of the Catholic Church, and celebrated this as evidence that the rest of the world should benefit from the Christian faith. Western civilisation was born, with its roots not simply in Greek or Roman learning, but in Egyptian wisdom, and a sense of the mysterious powers of water. The new power of the Christian church sat on pagan roots.

Thousands of years earlier, at Mari in Mesopotamia, a statue was erected near the entrance to the Court of the Palms. It showed a female figure with small fish apparently swimming up her body. When water is hand-pumped into hidden pipe work within the figure, it magically appears through tiny holes and cascades down the statue, making the fish look alive. The trickery of civilisation took on a whole new scale in 17th century Rome. Innocent x celebrated each August by opening the sluice gates and flooding the Piazza Navona. A lake would appear, at the centre of which the great obelisk arose, supported by the river gods. Power flowed on, wet and forceful, from the spring of the ancient world to the technology of the modern.

CHAPTER NINE

Taming the Rhine

OUR STORY NOT ONLY MEANDERS, it often becomes quite murky. There is no better example of this than in 19th century Europe, where two contradictory ideas flourish. On the one hand, the various offices of government are pulled together in the quest to drain land and make the rivers great arteries of commerce. On the other, people begin to romanticise the country and recognise the devil in the detail of water control. The seeds of the current debate between global development and environmental awareness were sown in the marshes along the Rhine.

Out of the Rhine would flow a paradox of civilisation. On the one hand came the idea that it was the purpose of government to control nature for man's advantage, while on the other a reverence grew of the natural world as something divine. Just as Germanic engineers were changing the great river from a meandering flow with straight lines and neat cuts, so the intellectuals of Germany were beginning to conceive of the green movement, of the majesty of untamed nature and the bounty of the earth. The artists and scientists may have differed in purpose, but they shared an obsession with the great river.

Here is the story of two men, both products of the age, and visionaries, but who imagined very different worlds; the story of the engineer Johann Tulla and the writer Johann Goethe. Between them they created the reality and cultural identity of modern Germany, and set the model for much of the modern world.

Imagine the countryside of northern Europe. The picture that comes to mind is an invented one. It has been constructed over the last 200 years. In the 18th century Europeans began a huge undertaking. They converted vast stretches of the continent into economically viable land. This meant diverting rivers, draining bogs, chopping down forests and changing the shape of the soil. Today's 'countryside' is a compromise between man's efforts and natural forces.

This was done in a haphazard way. There was no central co-ordination, rather a shared motive in turning a buck from agricultural land, or industrial processes. Yet there were leaders to this movement. The continent was coming out of centuries of bloodshed and disease. The Enlightenment, born in great cities such as Edinburgh, offered a more rational view of the world, which encouraged the idea that alternatives existed to war, and that science could conquer illness. It also re-inforced the idea that humans were the superior animal, and that the earth was for humans to govern. Western civilisation fostered the idea of the human at the centre of affairs.

If man was supreme, and scientific rationalism superior to the wild, then it made sense that nature was the enemy. The first and greatest threat came from uncontrolled water. Thus, our modern landscape is a result of 'war' waged by man on the territory of the north, the wet, river-riddled north that never suffered from the water shortages that produced the classical civilisations.

'Let us learn to wage war with the elements, not with our own kind.' James Dunbar's rallying cry of 1780 sums up the mood of the time[37]. The city, born to serve the country with labour in Mesopotamia, had become master of the land beyond its walls. No longer would urban dwellers see the rural stretches as the true home of man, but instead as a challenge, to be controlled and organised as neatly as a city. This idea lodged in European thinking and would be carried to the New World by European immigrants, who would battle against the American wilderness.

The Rhine runs from lower Germany, through France, to the North Sea and so the British coast. It had long been a trade route, as well as the source of fish, water and myth for countless communities close to its winding path. The energised Europe of the 18th century saw the many bends in the Rhine as a challenge. If more trade, travelling more quickly, was to be done, then the river had to be fixed.

To think of the Rhine of 200 years ago, it is perhaps better to imagine something closer to a whole country, with all the variety of that, which slinks and gushes across the continent, changing its borders according to rain and sun. Our modern image of a narrow river, contained in the same bed, does no justice to the old Rhine.

As the Thirty Years war draws to a close in 1648, the German landowners look at the marshy land and think of how they can improve it. The miracles of the Dutch in holding back the sea and draining the land are known, so people look to Holland for water expertise. And the Dutch are more than willing to help. They travelled beyond their borders, with the aim of creating 'new Hollands' – the boggy continent presented a challenge to the practical, hardworking puritan minds, and they were more than pleased to 'save' the land. In so doing they were taking land back from a kind of sinful indolence and making it virtuous through drainage.

Engineers such as Jan Leeghwater, who had drained Lake Beemster in Holland, and Cornelius Vermuyden, who had a mixed fortune trying to drain stretches of southern England, came to Germany and set about turning damp land to its godly purpose of usefulness. The fens close to Berlin were drained, then the engineers were despatched to the Oderbruch, a notorious swampy flood plain that had been dyked with only intermittent success.

The divine mission of draining land had a practical purpose. It is important to remember that life in Europe at the time was pretty miserable. Child mortality is running at about 50 per cent. The population is only just beginning to recover from years of plague. Disease is common. Despite the abundance of water, famine is also present as crops fail in the bitterly cold winters and wet summers. Controlling the water was thought to control disease, apparently spread by the smelly miasma of fog and damp that rose from stagnant water. Better drainage might also protect the crops.

The Oderbruch became a kind of laboratory for German water engineering – by trial and error, they set about taming it. In 1736, dykes and drains that had been built by the Prussian military proved too effective. They directed water upstream, so that the whole volume swelled and broke the defences. The floods that burst through were as devastating as if a major dam were to burst nowadays. Whole communities were wiped out.

The historian David Blackbourn is fascinating on the role of the Rhine in the formation of Germany and European attitudes to identity and the countryside. He argues that this wasn't a 'war with the elements', but an actual war. The king was in charge of the great drainage schemes of the late 18th century on lands. Much like Roman emperors before him using soldiers to build aqueducts, Frederick the Great used his troops to drain the water from the Oder Marshes around 1780.

The success of the scheme appeared to be new farmland, but Blackbourn says it also meant Frederick had acquired new territory to govern that was easy to protect. So when Frederick says, 'Here I have conquered a province peacefully', the king was in tune with the times.

Contemporaries referred to Frederick II's bid to control the Oder as a 'silently conducted seven years war' – it used all the resources of the state, and exhausted the army. But Frederick believed he had done something better than just win a new battle. He had advanced civilisation through dyke and drain: 'Whoever improves the soil, cultivates land lying waste and drains swamps, is making conquests from barbarism'. To him, and others, the view was that people who lived in the fens and marsh were backward, crude and diseased. They were 'barbarians'. The process of controlling the Rhine's watery excess was about civilising the people, and that meant civilising the land. The link between civilisation and manipulation of the watercourse was set, and would remain a key feature of western society.

The Prussians were delighted with their achievement, one writer speaking excitedly of a landscape of 'wealth and almost Dutch cleanliness'. The wealth was in part because an organised acre is one best suited to farming, and the task of growing crops was getting scientific in this enlightened age. The 'cleanliness' may in part have meant the neatness of the view, but also a reference to the reduction in disease. Malaria, though as yet unidentified, was harboured in the marshes – the infection rate must have fallen as the dry land spread.

With all this enthusiasm welling up, and the sense of divine purpose glossing the practical success, attention turned to the Rhine. It seemed obvious to the age that this was a resource to be put to the uses of man. 'A river is a road that moves' said Blaise Pascal, and the Rhine would be made into the first transcontinental superhighway. The haphazard amendments to the river that had occurred over hundreds of years would be replaced by a single, co-ordinated scheme. If this was a war against the elements, then the Oder had been a mere skirmish next to the battle that would commence. And, like all such ventures, there was a general to command the forces of man. Johann Gottfried Tulla was the fantastically self-assured engineer who saw the battle through, and his commemorative stone would remember him as 'The man who tamed the wild Rhine'.

Western culture has eulogised the conquest of the American West by European settlers. This is a result of a national psychology in the USA of breaking new frontiers, of the lone man facing up to the challenges of life, and has been helped by a film industry that spun endless simple tales from this material. Europeans imagine they have no such equivalent but a conquest occurred in Europe – the people fought the watery edges and pushed them back, until the rivers were contained and the marshes dry and the landscape built to suit the world's first experiment in industrialised farming and mass-participation industry.

Tulla saw the Rhine as the great challenge of the age, a puzzle to be solved for the benefit of all. His life would be spent immersed in the detail of the twisting river. Suitably, his family were originally Dutch. The clever Johann began his education in Protestant theology before studying science, and then engineering. It's as if his life mirrored the intellectual drift of the age. At the age of 34, in 1804, he was appointed chief military engineer to the river. In 1809, he set out his plan to modify the Rhine, which evolved into 'The Principles According to Which Future Work on the Rhine Should Be Conducted', published in 1812.

Tulla had drawn together a thousand bends and pools, and delivered a line. As if the Enlightenment spirit of rationalism had leapt from the mind and the pen and become real in the land, an actual place of enlightenment, where barbarians would literally be brought out of the murky bog and made to see the light of civilisation.

While Tulla was straightening nature out, his contemporary Johann Goethe was wrestling with the twists and knots of myth. The academic and dramatist retold the tale of Faust, and his version came to a conclusion that would have horrified the Enlightenment engineers. As Goethe drew together the shifting tales and themes of his work, he concluded that controlling water was the devil's work, and the natural river was closer to the divine.

Starting in the 1770s, Goethe took 60 years to build his narrative flow. The first part was published in 1808. The second part came off the presses after his death, in 1832. There is lively scholarly debate about how best to interpret the whole – is it unified, or a gathering of ideas over a lifetime? Everyone knows the idea of Faust – a man sells his soul to the Devil, and then tries to renege on the deal. Goethe's version is far less clear. His plot isn't simple. It's as if he is trying to tell two tales at once, or more. There is a love interest for the Catholic country girl Gretchen, which has little to do with the pact with the Devil. It's not clear if Mephistopheles is the Devil, a representation of the Earth spirit – an embodiment of the green man cult found across northern Europe – or a symbol for rationalism and cynicism. Odd as this may seem for such a well-known tale, it is vague what the exact deal is between Faust and Mephistopheles.

Goethe himself wrestled with the vagueness of it all. It seemed to be a eulogy to love and wildness and nature, but strapped to a plot that couldn't make up its mind. In various letters to friends in 1797 and 1798 he wrote of the 'misty and murky path' of *Faust* and of a 'barbarian production' which 'should appeal to a vast northern public'. In a poem of the time called 'Valediction', he wrote:

> Enough! Farewell now to the limitations
> Of this barbarian world of incantations!

This is why *Faust* is such an important work – it reflects perfectly the muddled, unstructured emergence of northern identity. Unlike all previous civilisations, the idea of being northern wasn't controlled within an autocratic political structure. Individuals were at liberty, not dependent on the all-powerful élite who controlled the water. Thus 'western civilisation' emerges, and has a self-image very different to previous ones. The 'enemy' is not drought, or thirst, so the task is not the endless labour of controlling the canals and ditches. The 'enemy' for those who wished to control the north is nature as a whole, yet this same nature is, for the poets and myth-makers, the soul of land. This is how European identity can combine an astonishing rush to industrialisation, while imagining it exists within rolling hills, forests and babbling brooks.

As Goethe's play is performed across Germany to rapturous audiences, Tulla is unveiling his great drama, the taming of the Rhine. He proposes to change 354 km of river between Basel and Worms. He does this with the great rational slogan of 'no river or stream ... needs more than one bed'. The river, he says, will be, 'where practicable, a straight line'.

The labour for this great endeavour comes from the army and from paid employees. Where once you needed slaves for such works, in 19th century Europe you need soldiers and navvies – the state couldn't force people to work on the schemes, but had to pay them. If this marks one key change between modern western civilisation and previous models in Rome, Athens, Egypt or beyond, then it shouldn't overshadow another crucial difference.

In order to get this done, Tulla required the co-ordination of all offices of state. We think of united Germany emerging out of wars and the unifying purpose of Bismarck. But a crucial factor was the modification of the Rhine. As the historian David Blackbourne puts it:

> Tulla's project meant, in effect, that the whole apparatus of state was mobilised. The departments that were called on for opinions and action included Foreign Affairs, Finance, Interior, Forestry and Mines, Domains, Waterways and Roads. But if the Rhine project reflected a larger pattern of state-building, it also contributed to that process. It was an undertaking that advocates hoped would integrate the new state along its major artery.

The bureaucratic problem was that Europe in 1808 was a place of many different authorities. The nation state as we imagine it had yet to be born. To get the work done numerous treaties and deals had to be brokered, interrupted by the Napoleonic Wars, and not completed until 1840.

Of the 354 km that Tulla worked on, he shortened the river's bed down to 273 km, making dozens of cuts and removing over 2000 islands. Between Basel and Strasbourg alone a thousand million square metres of island or peninsula were excavated, 240 km of main dyke constructed, out of five million cubic metres of material. This was backbreaking work – it took seven years to make one cut at Mechtersheim, begun in 1837 and finished in 1844.

Labourers and soldiers were making channels as wide as 24 metres across, deep enough to carry the full flow of water. Many villages feared the work would increase the likelihood of floods – a prescient worry, but one dismissed by the authoritarian Tulla. He was on a mission. To understand his purpose and works, he believed, depended on 'whether the active agents are more or less enlightened and moral'. Again, control of nature, and morality are connected. To Tulla, the immorality was close to stupidity: 'As a rule I consider it to be largely wasted effort to try to educate people on things that lie beyond their area of competence.'

Arrogant and inspired, Tulla oversaw an amazing achievement. Working against the flow of disjointed, warring Europe, of a ramshackle state structure, against engineers who said it could not be done, and Rhine-dwellers who wished it would not be done, he built a new artery which would carry the trade of a new civilisation – the west. He died in 1828, a proven man.

Four years later, in 1832, Goethe died. He had stipulated that part two of *Faust* should not be published until he was gone. Perhaps he never fully reconciled himself to the success of a play which was so confusing – so 'barbarian'. If he was seduced by the idea of classicism, of being a writer to match Homer or Ovid, then it must have sat uneasily to realise he was a champion of the north, of an unrecognised tradition. One hopes he realised that he had created a defining work of art for that emerging tradition, that he was as great as any southern author.

Given the distance of 24 years since part one had been published, the second episode of this tragedy has a different feel and a greater sense of narrative drive. Mephistopheles has made Faust into a powerful creature. The deal appears to be that no matter how much the devil provides, Faust will not become complacent and take the wonders of the world for granted.

In looking for a great metaphor Goethe picked the technology that was apparently transforming the continent – water control. He was fascinated by proposed canal schemes at Suez and Panama. He was looking for the next great wonder of a technology championed by the Dutch and then driving change throughout Europe. So Faust declares that he wishes to hold back the sea and to drain the land;

> A marshland flanks the mountain-side,
> Infecting all that we have gained;
> Our gain would reach its greatest pride
> If all this noisome bog were drained.
> I work that millions may possess this space,
> If not secure, a free and active race.
> Here man and beast, in green and fertile fields,
> will know the joys that new-won region yields,
> will settle on the firm slopes of a hill
> Raise by a bold and zealous people's skill.
> A paradise our closed-in land provides,
> Though to its margin rage the blustering tides;
> When they eat through, in fierce devouring flood,
> All swiftly join to make the damage good.
> Ay, in this thought I place my faith unswerving,
> Here wisdom speaks its final word and true,
> None is of freedom or of life deserving
> Unless he daily conquers it anew.
> With dangers thus begirt, defying fears,
> Childhood, youth, age shall strive through strenuous years.

> Such busy, teeming throngs I long to see,
> Standing on freedom's soil, a people free.

Goethe packs into one speech the essence of north Europeanism. There is the hard labour at controlling the water, which bequeaths a legacy of 'paradise' where good health reigns, people are free yet work together to a common aim. It is the struggle against water which forces Europeans to reconcile political structures with social aims – 'none is of freedom or of life deserving' unless he joins in the work at digging dykes and draining bogs. If all previous civilisations relied on coercion, slavery and dictatorship to achieve water control, and so their glory, then Europe would do it by linking the rights of the individual to the responsibility to hold back floods.

The consequence of this 'war' on nature was that the rate of water extraction and manipulation increased dramatically. The rivers weren't the prize for victory in this conflict. The trophies of war were bigger cities, better trade and an environment less vulnerable to flood or drought, at least in the short term. The 'conquest' of the Rhine set in train a process of water control which would allow for the massive expansion in civilisation. No one stopped to think that one day the water might be less biddable.

CHAPTER TEN

Health

THE GROWTH OF THE CITY is only possible if the pipes work. Up to a certain population size, people can live close together and rely on natural springs for drinking, and open holes for sewage. Much larger than that, and you have to become a master of piping clean water in, and dirty water out, if your metropolis is to thrive. Urban historians call it the leap from 'pit to pipe'. This may seem like a grubby, quotidian task, but without it the way mankind has developed on the planet would have been impossible.

Throughout history, cities had been limited by their access to water. A critical mass was reached when too many people were dependent on the same water supply. Jericho's population of a few hundred relied on one spring. Irrigation is developed, allowing for more food but also for the supply of fresh drinking water to the city, and Ur in Mesopotamia grows to a population of around 50,000. This is the first great advance in the potential size of urban living. Controlled water allows for bigger cities.

One idea about the city is that it is a place of sin and disease while the countryside is clean and healthy. Yet one of civilisation's qualities is cleanliness – as we have seen, Rome went to great efforts to build public baths; the steaming naked skin of the citizenry proof that Romans were the best people in the known world. It appears to be a contradiction: powerful people build cities, which are dirty, but need to keep clean to prove their superiority.

When cities are small, it is relatively easy to overcome this problem: the rich get clean and the poor stay dirty. Rome's baths were for the upper classes, not the peasantry or slaves. Providing enough water so that everyone could lounge in a hot pool would have been a mammoth task, and for little benefit; clean noblemen might be proof of civilisation, but clean slaves are a waste of time and money. The public fountains of the Roman Empire provided quite enough washing water, in the eyes of the rulers.

However, when cities get big, cleanliness is no longer a matter of luxury. It becomes a necessity. Dirt kills, and nothing slays people in quite such numbers as dirty water. This is why the expansion of the city is also the history of man discovering how to eradicate disease and dispose of filth. In doing so, he learns how to control or remove the single greatest threat to human health in 'civilised' society.

Life is a hazardous thing, and people will die from disease and infection no matter what, but the pattern of illness and mortality changes when mankind learns about agriculture and water control. With civilisation come epidemics. Diseases can spread from person to person, and from domesticated species to humans, and infect people at a rate that makes it hard to control. When humans stopped travelling and settled down, they entered a world of invisible enemies.

Ancient irrigated agriculture means standing in the water for a long time. Whether unblocking a ditch, ensuring the water goes evenly to each plant or digging new channels, the workers' feet would be damp. The water itself wasn't moving at any speed, but stood still in small puddles on the soil. These are ideal conditions for parasites that penetrate the skin and get into the blood stream. Schistosoma is one parasite, a fork-tailed blood fluke. Bilharzia or schistosomiasis will follow: when the worm is within your body, your mind and strength will deteriorate. Water control and civilisation literally sent people mad with irritation from such disease. Forensic analysis of Egyptian mummies has shown calcified blood fluke eggs in liver and kidney tissue, proving that ancient Egyptians were killed by schistosomiasis. It is reasonable to assume that the people of Mesopotamia, Mohenjo-daro and the Yellow River suffered similar fates.

We don't know if these civilisations connected working in the fields with the diseases. It would have been hard to make the link, given that the city dwellers were equally vulnerable to illness. The blessing of irrigation was a surplus of food, which was stored in granaries. These were incubators of insects and bacteria, quite apart from offering a warm home to rats, whose faeces would have been a common element in the people's daily bread. City life was bad for the health. Little wonder people had a lot of children; they were trying to outpace the death rate.

We can't tell if ancient civilisations made the connection between irrigation and disease, but it requires no more than a nose to link city life with bad smells. Hunter-gathering man could defecate a few paces away from camp – his only concern was to not sleep downwind. Small settlements around early agriculture would have easily coped with a drop latrine – a hole in the ground. In fact, simply covering faeces with soil would have sufficed. But when people are close together, and open countryside is more than a few paces away, toilet habits become everyone's concern.

We know early civilisations attempted to rid the streets of human and animal dirt. At Mohenjo-daro there were drop toilets in the homes of the rich, flushed out with used bath water. This 'brown' water, as it is now referred to in sewage circles, was sent to a cesspit which itself emptied into a large main sewer. This tunnel was big enough for a man to walk through, suggesting sewage was a serious matter that required a big investment. The sewer then drained into a river. The principles of city sewers were known and executed 5,000 years ago.

The reason for building a sewer is that the human and animal excrement would otherwise clog up the streets. They were intended to take the stench away. There are no records suggesting early civilisations built sewers for medical health. The idea that disease might spring from dirty water is only 150 years old. There was a broad sense that clean water was good. Greek medicine believed this and accounts for why the aqueducts into Athens were covered, so that the water was 'pure' for drinking. The public baths of the city were part of the gymnasiums, where people exercised, so there was a cultural link between being healthy and washing.

Vitruvius, who wrote about the principles of Roman architecture in 27 BC, describes the sanitary necessity of good water supplies. He writes that any architect must know '... the characters of the atmosphere, of localities (wholesome or pestilential), of water supply. For apart from these considerations, no dwelling can be regarded as healthy.'

This explains why there were 144 public loos, over 800 baths and an elaborate system of clean water delivery and waste water removal at the time Sextus Julius Frontinus wrote *De Aquis Urbis Romae* around 100 AD.

Rome's large sewer was the Cloaca Maxima. This was originally dug around 500 BC in order to drain the marshy land on which the city would be built. It predates the first aqueduct by 200 years. As Rome got richer, the Cloaca Maxima was covered, though this didn't diminish its scale. Emperor Agrippa is alleged to have sailed along the channel, which was 900 metres in length, up to 3.2 metres wide and, at points, over four metres high. It was such a marvel, making the growth of Rome possible, that it was awarded its own god, Cloacina.

While other sewer pipes fed into the Cloaca Maxima, running under the paving stones of main streets, very few houses were directly connected to the system. The reason is simple. At times of flooding, the pipes would back-up and overflow, dumping raw waste in a city home.

Ancient sewers don't mean old cities were clean. Rome's streets were littered with excrement. The authorities organised gangs of waste collectors, but that probably had only a marginal effect. Our cultural memory of the city as a place of dirt, of physical and moral decay, is an expression of fact. Urban living presented a new problem to humans – how to clean up.

The problem had not changed a thousand years after Rome's glory. Early medieval Europe was troubled by the same stench and disease that plagued previous societies. Records show towns such as Bruges, Paris and Milan had city laws governing the sewers. The main waste disposal system was either a direct route in the local river, which was also the source of drinking water and fresh fish, or into cesspits. The principle of the cesspit is that the heavy waste settles on the bottom while the liquid matter drains away – in theory, a cesspit should be like a compost heap. In practice, the liquid matter often doesn't drain, and is added to by rainfall and ground water. If the heavy waste remains wet and sludgy, it is a richer environment for disease and smells a lot worse. We should admire the city fathers who drew up the ordinances, but they were on a hiding to nothing.

London's statute book of city regulations would grow thick and run into volumes on the subject of sewage alone. In 1312 fishmongers were instructed to throw waste directly into the River Fleet or the Thames rather than dump it on the street. Two years later there is an ordinance attempting to control general pollution. The city was dependent on the tidal Thames to clean the human mess – Londoners worked on the principle that the ebb and flow of the river would take their faeces to the sea. Indeed, a person could enter a public latrine on London Bridge and dump straight into the murky waters below.

The alternative was to defecate over a cesspit. A man called Richard the Raker was doing just that in his private privy in 1326 when the floorboards snapped. He drowned in the slurry below. Had he survived, he would have needed no persuading that civilisation needed a better form of sewage control if it wanted to expand.

As it was, a more fearful, tragic lesson on the merits of urban sanitation was about to be dealt out to northern Europe. Between 1350 and 1450 the continent suffered wave after wave of plague. The disease was nurtured in the rank city sewers, carried by the rats lured to the mounds of human waste. It spread from the urban centres like puffball mushrooms bursting their spores over the land.

London suffered bouts of the plague in 1361, 1369, 1370, 1382, 1390, and 1407. There is no accurate way of assessing how many people died. The terrible human price it exacted could have been taken as a warning about the perils of city life – a dark echo of biblical tales from Gomorrah. But the appeal of city life, of the attributes of civilisation, was so great that mankind, as elsewhere, chose to pay the price.

The plague would return 200 years later. It is possible to be slightly more accurate in assessing mortality during this second period of bubonic infection. In 1563 17,500 Londoners were killed. There would be four more plagues, each more dreadful, the last in 1665 taking 80,000 lives. Europe would never rise from its damp fields and match the glories of previous civilisations, if it didn't overcome the problem of filthy cities.

The critical mass of how many people could be sustained within one area didn't change much over time. London in Henry VIII's rule, at the beginning of the 16th century, had a similar population to Ur's some 4,000 years previously. Europe's growing wealth and population boom required the city to accomodate more people. If the grand ideas of the modern state were developed in the Low Countries, Renaissance Italy and on the banks of the Rhine, much of the practical work occurred in Britain. Having taken the lead in the Agricultural Revolution, and then nurtured the Industrial Revolution, the British were the first people to face the problems of modernity. They would pioneer water control that would make our urban world possible.

'London, a city famous for wealth, commerce and plenty, and for every other kind of civility and politeness; but which abounds with such heaps of filth, as a savage would look on with amazement.'

That was Lord Tyrconnel's verdict on Britain's capital city in 1741. He makes a clear link between cleanliness and 'civility'. Yet the city is besmirched with 'heaps of filth' that would shock even a savage. The assumption is that the savage, bereft of civilisation's bounty, would still manage to keep things cleaner than Londoners. Imagine, Tyrconnel seems to say, one of the largest and most powerful cities in Europe should be dirty – as if this contradicted any claim to advancement.

The fact is the idea of civilisation as a clean place, a washed environment, was challenged on every city street in Britain. In 1759 the monarch,

George II, would describe the year as an 'annus mirabilis' – a miracle year given the nation's military success abroad and its intellectual and cultural confidence at home, all wrapped up in a booming economy. How could the nation call this a glorious achievement when everyone lived in stench and disease? The problem of how to create a clean city would grow more pressing as Britain, the first nation to experience industrialisation, found living standards dropping to 'savage' levels amid the new wealth.

British cities swelled to previously unimagined size. In the 19th century London had a population of over a million and Glasgow 250,000. People were drawn in by the promise of work and wealth, only to find themselves stepping back millennia in terms of their life expectancy. The average life span of a person living in Catal Huyuk around 7,000 BC was 20 years. For people in England's Black Country, one of the hotbeds of industrialisation around the city of Birmingham, the life expectancy was 17 years. In Glasgow and Manchester the average person would be dead before their 28th birthday.

Dr James Phillips Kay left Edinburgh for Manchester in 1827 and he described the awful quality of life in the English industrial town in his 1832 work 'The Moral and physical condition of the working classes employed in the cotton manufacture in Manchester'. He exploded any notion that city life, civilisation and cleanliness were one and the same. These were terrible slums, riddled with cholera, close to a river that must have stunk with putrid matter. It is estimated that in the 1840s 70,000 tonnes of human faeces was being dumped in its flow.

London was just as bad. Sources refer to human excrement lining the banks of the Thames. In 1832 there was an outbreak of cholera that killed 32,000 people. Glasgow, the powerhouse of Scotland's industrialisation, was growing faster than any other European city. The Gorbals, a part of Glasgow close to the Clyde River, was at one point the most densely populated place on Earth, with humans crowded upon humans, a haven for disease, squalor and rat infestations.

In 1837 the general Register Office was formed. This meant statistical accounts on health are available, and the first report said poor sanitation cost 137 lives a day. 16,000 were killed by typhus a year. Between 1848 and 1854, the death toll from cholera was 250,000 and 15 per cent of children didn't survive infancy.

In 1842 the Secretary of the Poor Law Commission, Edwin Chadwick, would edit an influential paper for the House of Lords, entitled 'Report on the sanitary conditions of the labouring population of Great Britain'. One of his correspondents was a medical officer in Macclesfield, a town to the south of Manchester. He wrote:

> ... to these (34) houses are three privies uncovered, here little pools of water, with all kinds of offal, dead animals and vegetable matter are heaped together, a most foul and putrid mass, disgusting to the sight and offensive to the smell; the fumes of contagion spreads periodically itself in the neighbourhood, and produces different types of fever and disorder of the stomach and bowels. The people inhabiting these abodes are pale and unhealthy, and in one house in particular are pale, bloated, and rickety.

A process that had begun in the early Middle Ages, of legislating for a cleaner city, continued. In 1840 the Select Committee on Health of Towns recommended a General Building Act. The aim was to end the phenomenon of back-to-back houses, which had the effect of people living in crowded spaces that were almost impossible to drain or plumb. The Committee also recommended a General Sewage and Building Act and a General Act to aid local water supply.

What resulted in 1848 was a milestone in public legislation. Up until that point individual cities were responsible for their water supply and sewerage and could address these in any way they saw fit. As the city fathers were by and large the same rich men who profited from the factories, there was little incentive for reform.

So when Britain's first Public Health Act became law in 1848, it struck a blow against these wealthy entrepreneurs. Corporate boroughs were given responsibility for drainage and water supply. What was important was that central standards were set for that supply. It now seems obvious that any utility company, indeed any service of commercial activity, would be governed by a single set of guidelines, but these bordered on the revolutionary in the eyes of industrialists, who fought them in a vigorous national campaign. They lost. Britain set in stone a basic principle of modernity, that people could expect uniform standards, and had in particular the right to receive clean water and have their waste taken away.

The authority to challenge these rich men and their cost-cutting ways was derived in part from the democratic system, and in part from God.

The link between the city, associated with the high ideals of civilisation, and squalor, offended the Victorian sensibility. Much as Lord Tyrconnel found a hundred years previously, the mix of city life and filth somehow didn't compute, as if it was against some fundamental law for the savage and the civilised to sit side-by-side.

However, the church carried sufficient weight to counter the voice of profit. And this is not only the established church but the broader sense of society which was permeated by religion and which stood in deference to faith. When the Health of Towns Association was set up in 1844, it took as its aim: 'To diffuse information as to the physical and moral evils that result from the present defective sewerage, drainage, supply of water, air, and light, and construction of dwelling houses...'

Dirt was a moral issue. It offended the muscular Christian of the Victorian age, so they connected cleanliness to godliness. In doing this, they merely continued a tradition of civilisations, linking the control of water to spiritual blessing. Washing was a route to the divine. As the historian Tristram Hunt puts it: 'The cities needed to be cleaned so the people could be cleansed.'[38]

This close connection between Christianity and sewerage is best displayed in the children's book *The Water Babies*. The author Charles Kingsley spins a tale of grubby working children falling into the canals and, as they roll in the water, plump like cherubim in a Renaissance painting, they become nicer people.

Kingsley clearly had an advanced sense of communicating a message, using nursery books alongside his main occupation of giving speeches about the ills of dirty streets. A Christian socialist, his views were summed up in the lecture 'Great Cities and Their Influence for Good and Evil' in which he said that a moral life could only come from the abolition of 'foul air, foul water, foul lodging'.

Much as the Romans had venerated their sewers by awarding it a God, so the Victorian Britons took the task of sanitation to be a godly one. Viscount Edrington of the Health of Towns Association said: 'The health, the tranquility, the morality, nay, the Christianity of the people of this country are nearly concerned in the sanitary conditions of these towns'. This explains why it was the Bishop of London who introduced the 1846 Public Baths and Wash houses Bill to the House of Lords. It was a great success.

In political terms, with great relevance to many of the water disputes today, this is an important episode. Linking the environment with moral health, and by association physical health, remains a key association. Arguments for and against major dam projects or the privatisation of water often turn on this issue.

If God wasn't enough to stir political action, then legislators had a more direct reason coming. The Palace of Westminster, which houses the British legislature, was built between 1840 and 1860, replacing fire-damaged and decayed buildings on the same site. Members of Parliament must have been quite delighted when they gained entry in 1858, thinking themselves very wise to have commissioned such a grand building. Any smugness was removed when the windows were opened. Sited on the banks of the Thames, it enjoyed the full richness of the city's stench. That year, with a hot summer, things were particularly bad. This prompted debates in chamber on what was dubbed 'The Great Stink'. The Tory MP Disraeli described the city as a 'stygian pool reeking with ineffable and intolerable horrors'. Something had to be done.

The decision was taken for London to copy Liverpool, which became the first industrialised city in the world to have a proper, purpose-built sewage system designed. It was drawn up by a Scot, James Newlands, in 1847. Sir Joseph Bazalgette was tasked with building London's sewers, and it took nearly 20 years, between 1858 and 1875.

Bazalgette used gravity and pumping stations, laid 82 miles of sewer, which carried 52 million gallons of waste to 14 miles beyond Tower Bridge, with the tide carrying the waste out to sea. So great was the achievement that both the Prince of Wales and the Archbishop of Canterbury helped in the opening ceremony. By getting rid of open sewers and stopping raw sewage going directly into the Thames, London was granted the key to becoming the world's first megacity. It suffered one last cholera epidemic in 1866 and by the turn of the century had a population of over a million people.

Modern politicians invoking a sense of Britain conjure up ideas of a leafy countryside where cricket is played and jolly locals drink pints of beer. This stands in stark contrast to the Victorian idea of the country. For them, the cities, which had been tamed by plumbing and sewers, were the pinnacle of civilisation. Nationalist rhetoric about what made Britain great centred on the city, its size and the diversity of people. If you wanted to know why Britain deserved its status as top global nation, and the huge empire it had acquired, then you only had to walk down one of London's grand streets. Now, it was supposed, Lord Tyrconnel's savage would be awed by the Christian cleanliness of it all.

No doubt the street life of London would have amazed most of the people on the planet but they, and future generations of humans, would be more impressed by the practical work of Dr John Snow. An anaesthesiologist by training, he had become fascinated by epidemiology. He plotted the occurrence of cholera outbreaks to try and fathom their cause from the data. The prevailing view was that the disease formed a 'miasma' in the atmosphere – they believed a strong smell would infect them.

Snow theorised it was not some mysterious air that infected people with Asiatic cholera, but contaminated water. He proved his point by studying the 1854 outbreak in London's Soho district. In the first three days of September, 127 people died in the area around Broad Street. By 10 September 500 people were dead. This was a very high mortality rate, and caused panic.

Dr Snow stayed calm and, from talking to many people, began to deduce that the Broad Street water pump was at the centre of the outbreak. On 7 September he studied a sample from the pump under a microscope and found it contained 'white, flocculent particles'. He convinced the local parish to remove the pump handle, and subsequently the infection stopped.

By investigating various cases that were outside Soho, he was able to establish the common link in infection was the pump. A Reverend Whitehead set about trying to disprove Snow's thesis, but found himself agreeing with him, and concluded that the nappies of a cholera-infected baby being washed out into a cesspool a few yards from the pump was the likely cause of the outbreak. However, a Board of Health enquiry some months later rejected Dr Snow's thesis, and little changed in Soho, as *The Builder* magazine was to report in 1855:

> Even in Broad-street it would appear that little has since been done... open cesspools are still to be seen... in spite of the late numerous deaths, we have all the materials for a fresh epidemic... In some [houses] the water-butts were in deep cellars, close to the undrained cesspool.

In 1855 Dr Snow dug into his own pocket and printed an updated version of his long-held theory under the title *On the Mode of Communication of Cholera*. Though we cannot point to one moment when the establishment realised he was right, the rigour of his work contributed to the growing campaign for reform of urban sanitation. Sir Joseph Bazalgette's sewers are indebted to John Snow's wisdom. Dr Snow's link between clean water and good health, and contaminated water and cholera, remains one of the single greatest contributions to human health on the planet. Clean drinking water is a greater influence on disease than all the inoculations and vaccines put together.

CHAPTER ELEVEN

Coal and Cotton

COAL AND COTTON BOTH appear profoundly dry. The scuttle filled from the old shed may glisten with the evening's rain on its lumps, and cotton may cling to a swimmer's body after an impulsive dip, but both will dry out quickly – that is why they are valuable. As such, they seem odd commodities to discuss in a book on water. Yet the widespread use of them accelerated the approach of peak water.

Physical power drives the spread of civilisation and progress. Brute force advances the benefits of humankind. Controlled energy determines how much land you can put under tillage, how high and strong you can build your homes, and how determined you can be in resisting your enemies. Most importantly, to our modern civilisation, the control of power determines your degree of industrialisation, and the strength of your economy.

The control of water gave rise to the first agricultural revolution and led to the first industrial revolution. Agricultural yield, the amount of food gathered from one field, leapt in quantity with irrigation. To this day, irrigated land is anything up to four times more productive than non-irrigated land. The world is fed thanks to the irrigation ditch. The 'Agricultural Revolution' usually refers to changes in farming in England from 1,500 AD on, but the control of water was just as revolutionary, and occured millenia before.

The control of water also led to the rise of industry. With water, man could harness much more power than either by his own muscles or those of an ox. The mill wheel, grinding grain into flour, driving saws in timber yards or pumping air into blast furnaces, would liberate man from the limits of flesh. He would see for the first time that his might was limited by his imagination, not his physical frame.

The first evidence of water mills occurs in ancient Greece. Like many innovations, other civilisations were not far behind, with the Chinese and Japanese also realising that a wheel, lying flat, with one edge in a flow of water, will rotate. This system works, but very inefficiently. The weight of the wheel is a problem, requiring a powerful flow of water, and its proximity to the water means whatever is being milled may get wet.

The Romans take the water wheel a step further. They move the wheel upright, and develop gears, so that the power of the rotating wheel can be directed to drive a horizontal millstone up to five times faster than the flow of water driving the mill. Having a vertical wheel seems obvious now, and the existence of upright wheels to lift water from irrigation pools to ditches, as used in Egypt from around 1,500 BC, adds to the sense of puzzlement that this was not developed sooner.

Whatever the origins of the Romans' eureka moment with the water wheel, their innovation released a whole new world of energy. Without vertical lifting wheels, the gold mining industry, so important to the Roman economy, would not have succeeded. At Rio Tinto in Spain, they constructed a series of wheels capable of lifting water 30 metres out of the mine.

There is some debate about the Roman attachment to the waterwheel. Some archaeologists argue that, given the limited surviving evidence of water mills, they were not widely used. Instead, a horizontal millstone driven by an animal such as an ass or ox, was the main source of Rome's flour. The other view is that the evidence has been destroyed by time and subsequent human developments, but the water mill was important to Roman production. This is what the leading expert on Roman water use, A. Trevor Hodge, believes.

He cites Vitruvius, who wrote a lengthy and highly regarded book on Roman architecture and aqueduct design, who describes the use of vertical wheels. In particular, these are 'over-shot' designs in that the water falls on the top of the wheel and uses gravity to drive it round. The alternative is the under-shot wheel, which is dipped into the flow and uses the current to drive it round. Trevor Hodge also suggests that it may have been commonplace to set an under-shot wheel in an aqueduct channel. When you have gone to all that effort creating a controlled, steady flow of water, this seems like a sensible thing to do. However, much like the puzzle of why the vertical wheel wasn't developed sooner, it is idle arrogance to wonder at the technological mistakes of the past.

At Barbegal, north east of Arles in modern France, are remains of Rome's prowess with the water wheel. Two parallel rows of eight water wheels drove grain milling on a huge scale. It must have required large numbers of people to operate. Given the power source, water, was free and constant, perhaps there was even the temptation to run the mill from dawn till dusk, on something approximating a double shift. This vast operation was designed to feed the 80,000 soldiers, officials and relatives based at the military camp of Arles.

The first century AD poet Antipater of Thessalonica captures the paradox of how the labour-saving device of the mill in fact involves great effort to operate:

> Cease from grinding, ye women of the mill; sleep late even if the crowing cock announces the dawn. For Demeter has ordered the Nymphs to perform the work of your hands, and they, leaping down on top of the wheel, turn its axle, which with its revolving spokes, turns the heavy concave Nysirian millstones. We taste again the joys of primitive life, learning to feast on products of Demeter, without labour.[39]

The ingenuity of the Romans lived on. The *Domesday Book*, recording the assets of England in 1086, numbers over 5,000 water mills. Most of these were the less efficient undershoot method. Two hundred years later Cistercian monks built water mills into their monasteries. They favoured the overshoot style. They also advanced the design of the mill. By attaching a cam to the axle, they could power a wooden hammer's rise and fall, which was used for pounding newly woven wool. This process, known as fulling, was vital in creating England's first major cash crop, wool, and was the first example of the industrialisation of that industry. The wealth generated would pay for the expansion of Cistercian monasteries throughout the British Isles.

The people of the Low Countries were using windmills as pumps, thus draining large tracts of reclaimed land from the sea. The 'Garden of England', Kent, was where iron manufacture began, thanks to the use of water mills. This dirty but vital process would only move north when coal overtook water as the main source of power. The picturesque ponds of Kent villages were once the source of England's emerging industry.

Northern Europeans added sophisticated gearing and attachments to the water wheel. Apart from the hammer already described, the people of England and the continent were able to drive furnace bellows, saws, weaving looms, and to operate several functions at once. Mankind was now set on the course of industrialisation. In the 1500s a traveller to the Harz mountains of Germany would have come across a scene of hellish proportions. Here 225 waterwheels operated in a single system, powered by water delivered from a dam by an intricate and lengthy network of canals. The control of water allowed this mill to provide about one thousand horsepower.

Traditionally, the Industrial Revolution is dated to the late 18th century, when the cotton mills of northern England throb into action. The argument can be made persuasively that this is the second industrial revolution. The first is the gradual process over two and half thousand years, of learning about the wheel, of setting it vertically rather than horizontally, and then supplying it with gears.

The manufacturing economy of Britain at the beginning of the 1700s was based on home production, and output was limited. In the next 150 years, the second industrial revolution would see the invention of the factory, the massive increase in controlled power, and the emergence of new social order geared around manufacturing, which would set the model for all industrial societies.

As wood supplies were exhausted, people switched to coal for energy.

This meant more coal had to be dug up, which meant mines had to be sunk deeper. The problem with digging a hole in a wet country is that it fills with water – much as a sandcastle built below the high tide line will find its moat naturally filling up. You need to get the water out to dig the coal. The Romans used elaborate systems of wheels to lift water from their gold mines, but these required a lot of space within the ground. The narrow shaft of a coal mine meant wheels wouldn't do.

Instead, horses were harnessed up to verticle axles known as capstans. As they walked round, water was pumped up. This was slow and inefficient. In 1698 Thomas Savery made real an idea first conceived of by the Dutchman, Huugyens. Savery found a way of replacing the horse with an engine. He used condensed steam to form a vacuum within a pumping chamber, which sucked the water up. It required a lot of energy to operate, and removed relatively little water. In 1712 Samuel Newcomen designed a better engine. It used steam to drive a pump.

Newcomen's engine inaugurated the second industrial revolution. It was now possible to get enough coal, at a reasonable cost, for wood to be replaced as the main source of energy. However this story isn't about coal alone. At the same time, the cloth industry took an equally important step forward.

While international trade had brought England into contact with cotton, it was the Dutch who kick-started the industry. They came to settle in East Anglia, partly driven by the request of landowners who wanted to turn their flat, boggy acres into dry, fertile fields. Engineers from the Low Countries came to repeat their trick on the fens. Many drainage schemes proved laughably inefficient and expensive, but while the ditches were being dug, the incomers were also trading in fustian, a mix of cotton and linen. The Dutch, ahead in terms of industrial development, imported their looms to aid this work.

It would seem we get the phrase for incomprehensible noise from the cotton industry, as the looms were known as 'Double Dutch looms'. We can imagine the clatter of wood and cloth being compared to the native tongue of the Low Countries people. As it turned out, England and Britain would rapidly become fluent in the production of cotton.

Production of all clothes up to the 17th century was done in homes and small workshops – nothing bigger than a room. Wealthy entrepreneurs in search of new riches would employ these home workers to produce orders. People didn't need to live and work on top of each other. This would change in 1704.

In that year Thomas Cotchett leased the rights to the water in the Derwent River, in Derbyshire. Two years later he had 16 'Double Dutch looms' being powered by a water wheel for his silk mill. Bankruptcy foiled Cotchett's dreams of wealth, but John and Thomas Lombes took over the mill, and made a fortune from the spun cloth. The Lombes set up similar mills elsewhere in England, and they set a template for progress. The age of the home worker was over – the factory was about to arrive.

James Hargreaves invented the spinning jenny, which increased the amount of cloth that could be spun by one person. In 1769 Richard Arkwright took this to the next stage, making much larger jennies. These needed more power. Arkwright began with horses in harness to a capstan, but soon the fresh water spilling off the Pennines in England's cloth-making county of Lancashire were using over-shot water wheels. Whatever the power source, the new mills needed several people to work them. Having seen the crowd of 300 people labouring in Lombes' Derby mill, Arkwright put his workers in one building and made them do a variety of tasks under one roof.

So coal and cotton production both advance considerably when fresh water is controlled to their ends. More coal can be dug up, providing more energy for industry. More cotton can be produced, allowing merchants to expand their markets. The conditions are in place for Britain's power to rest not on its seafaring prowess alone, but on industry. If the drainage of the Low Countries helped to create the secular, property-owning democracy, ordered by law, then the mastery of water in Britain helped to form the modern industrial state, where the main economic driver was not agricultural production and sea-trade, but manufacturing.

When factory owners found water wheels too weak for the demands of the cotton market, they looked elsewhere. The answer to their power needs was found in coal mines. James Watt had taken Newcomen's engine and significantly improved it, by adding a second chamber. In simple terms, the steam was compressed, thus creating more pressure, and more power. Watt did this so that miners could go deeper. His invention changed the world forever.

Cotton mill owners took Watt's engine and found they could produce far more than under the old systems. Factories got bigger, and markets expanded, and more factories were set up, requiring more people to work in them, and more coal to fuel the steam engines.

The model of production for cotton, of household mills and a scattered workforce, was switched to one of factories and coal-powered steam engines. This drove the expansion of the cities. More was to come. To connect these urban centres of manufacturing, a transport system was needed. Roads were rough and the horse still the engine of choice. This was insufficient for the scale of coal and cotton that needed to be moved. The first solution was to build canals. Engineers such as Thomas Telford set to work creating viaducts to match the wonders of the Roman Empire. Ancient aqueducts were designed to move water in a dry land. The channels of the new system were to move commerce, floating on the bed of a commodity that northern Europe had an excess of – water. Much as Mesopotamia had thrived from the ability to create an excess of food and to move it around by waterway, so the Industrial Revolution updated the model and made it work for the modern era.

Watt's design had potential beyond mine pumps and cotton mills. Richard Trevithick tweaked the engine for more power, and bolted it to a carriage on which he motored through London in 1803. Twenty years later George Stephenson thought it might work better if the carriage was on rails, like the tram lines used in mines, and so invented the steam-powered locomotive. The first railway line in the world was opened in 1821 between Darlington and Stockton. Eight years later a steam train beat a horse-powered carriage from Liverpool to Manchester, and from that point on the horse would gradually retire to race tracks and country estates.

While the great empires of the east had used water to establish a consistent food supply, the north used it to lubricate the growth of capital. In the east, water control meant simple social structures of a ruling élite and a labouring class. In the north, water helped create a more complex hierarchy, based on wealth and skill. This liberal society found water could drive the mills that kick-started major industries. This created factories, and the lure of urban living and higher wages to draw people from the countryside. It could move barges full of coal and cotton. Steam powered larger engines, pumps, trains and then ships.

This may seem a simple point, but it is important to understand that the industrial society that emerged out of Britain would be copied around the world. The democratic, capitalist structure of state was the template for others. This model could only have been made in a place where water is plentiful. It relied on the key substance being so available there was no incentive to put a price on its worth. While wood, then coal, were natural resources which were turned into commodities – things which could be traded, which would determine the wealth of nations, as Adam Smith so brilliantly summed it up – water was not. There were issues over water rights, but water escaped the grips of capitalism. There was not a market in water.

This suggests two things. The north European (and now north American) model of industrialised, liberal, capitalist society may be best suited to wet countries. Secondly, we are quite used to making naturally occurring things, such as coal, or oil, into assets. The argument that it would be unnatural to create a market for water is historically flimsy. The only factor that differentiates water from other commodities is that it is vital for life.

Before we embark on these bigger, more abstract ideas, there is a practical point that needs to be made. The cotton industry always relied on foreign markets. The poor in Manchester's slums weren't wrapped in soft cotton. The cloth was exported. Initially, it played a major part in the slave trade. Cotton, cheaply printed with bright patterns, was traded in Africa for slaves, who would be shipped to the West Indies and American colonies. It is a strange irony that the gay clothing associated with Africa is a vivid legacy of colonial times and barbaric practices.

Obviously the market was international in sourcing too, as you can't grow cotton in Lancashire. It became such a dominant crop because the British Empire expanded to areas like Egypt, Pakistan and India where cotton was a natural choice. To make the colonies financially productive – for the land grab to meet the aims of a capitalist society – raw materials were needed which had an existing market in Britain and Europe.

What this meant was that Britain exported to two of the world's great river basins, the Indus and the Nile, industrial scale irrigation to support cotton farming. If the British industrial model was to be copied, then so were its agricultural practices; the vast water engineering schemes embarked on became monuments to the empire. Britain set a pattern for water usage which would be emulated, and bequeath tracts of the world a dangerous legacy.

Where Indians had constructed irrigation, the new rulers would expand it. Where there was none, new schemes would be built. This was water as power, politics and money all wrapped into one – but the aim was not to subjugate the natives by shackling them to the endless needs of water management, but to turn arid land into fecund acres, capable of bringing greater wealth to the people of London, Manchester and Glasgow.

The first modifications of existing irrigation schemes in India began in 1817, and as elsewhere, small beginnings gave encouragement for grander designs. A famine of the late 1830s prompted larger schemes, such as the Ganges Canal, finished in 1857. This led to a massive attempt to irrigate a belt of India stretching from Delhi to Calcutta. All fired up with theories and formula devised by the engineers who had tamed the Rhine, Sir Proby Thomas Cautley created a huge network of canals and channels. Unfortunately, his gradients were too steep, forcing the water to run fast, eroding both the channels and the fields. Evidence points to a dramatic worsening of crop yields as a result of this scheme, though it appears to have ultimately improved. The trial and error required to correct this mistake – equivalent to building the motorways of Europe out of material too soft for cars – led to greater expertise in hydraulic engineering. This hard-won knowledge would allow Britain to go cotton-crazy in its colonies.

Explorers had long marvelled at the flow of the Indus, and wondered what use it could be as a trade route, or source of energy. The British Empire, millennia after the glories of Harappa, tried to tame this great river. Small irrigation schemes on the Ravi and Sutlej rivers in the early 19th century led to larger ones, such that three million hectares of the Punjab was irrigated by 1900. This work on tributaries encouraged a scheme tapping the Indus directly, which when finished in 1932 was the largest water scheme in the world.

Previous civilisations would have sighed with envy at the work done in the Indus basin. An arid land had been turned into a major commercial centre. Land that could only sustain a small population beforehand was now suitable for a huge workforce. Unlike early empires, this grand scheme wasn't to honour God, or kings, but money.

The water jackpot in what is now Pakistan inspired an irrigation boom across the sub-continent. Ground water was pumped up and farms bloomed. At the beginning of the 20th century, India alone had twice as much irrigated land than existed in the whole world a hundred years before. Irrigation was thus established as a vital element of the emerging industrial, globalised world. By 1947, when India and Pakistan would leave the ambiguous embrace of Britain's empire, their territories were the most water-controlled lands in the world, with grand canals and private pumps all contributing to 28 million hectares of irrigation.

The trick wasn't performed in India alone. The British Empire also controlled Egypt. Muhammad Ali had been the first to build dams on the delta of the Nile, inspired in part by an idea from Napoleon. The British expanded this policy, constructing the first Aswan Dam in 1902, and going on to enlarge it twice by 1934. This performed a great trick – the river of myth and magic was now the hardest working course in the world. Fields were watered with the ancient practice of basin irrigation, and with new pipes and sprinklers. The green strip of land that follows the water down to the sea was on year-round production.

The British Empire didn't change the world because of its military might alone, or just because of its industry and inventiveness. It changed the way water and land was manipulated for economic gain. It bequeathed to vast areas a legacy of irrigation and intensive farming. Much as the sun was destined to set on the empire, however, so the water would inevitably dry up.

CHAPTER TWELVE

Dams and Politics

A STRANGE THING OCCURRED when dams were built on the Rhine in the 19th and early 20th centuries. People not only celebrated their completion with festivals and grand ceremonies, but they would come back, every Sunday, and admire the great slabs of engineering. From the outset, the dam was a good day out. They were the new pyramids.

German identity was closely entwined with nature. The forest and rivers were part of the Germanic soul. Goethe's *Faust* is torn between the business of rational control and romantic desire, and the crowds on the dam would have known this yearning for a romantic, green alternative to their dirty cities.

All around, Europe was turning to its water for relief. People sought out the purifying waters of spa towns, where they would drink or bathe in mineral pools. This set the foundation for our modern tourism industry. It led to the rail companies building tracks to the spas, and thus the machinery of the modern state begins to turn its attention to the business of pleasure and happiness. For those who couldn't leave town, bottled mineral water was brought to them.

Europeans adopted a view of water with echoes of the Arabic/Islamic attitude. They came to admire the blue water, and to connect to the idea of green nature. The day-trippers on the dam were toying with their ancient past of pagan rituals and woodland frolics. They were staring into the flow and looking back in time, to an age that wasn't regimented by the factory clock. This was innocent stuff – the immediate pleasure was from the fresh air and possibly a hot dog and ice cream for lunch. But they stood at the beginning of a particular phase in human history, when the business of controlling water would become highly political.

Water control has always been close to power. One academic argues that a particular kind of political structure grew out of water control. Karl Wittfogel could easily have been the name of one of the trippers to the dams. He was German and lucky enough to escape to America before the Second World War. In the 1950s he published a book that set the tone for water historians for decades to come. To read him now is to be flung back into the mid-20th century, when the world was viewed through strict ideological spectacles. The Marxist theory, and counter arguments, all look a bit out of focus now, the eyes blurring at concepts once so essential to the world, which are barely mentioned today.

Wittfogel made an important assertion. As a graduate, he began his career immersed in Marxist theory. Mid-career, he lost faith and switched allegiance to liberalism. His doubts were provoked by a paradox – Marx argued that in all societies the workers would inevitably revolt and assume control over the means of production. While there was plenty of evidence of revolution in Europe, there was barely any in Asia. For Marx to be right, it had to be universal, yet it apparently wasn't.

Wittfogel answered this paradox by refuting Marxism. He said societies weren't all the same, and wouldn't lead to revolution, and he offered as his evidence what he called the 'hydraulic societies' of the east. These were ones that had been formed around the control of water. He pointed to the places where civilisation began, and the societies that developed out of those roots, and said hydraulic societies has no incentive to revolt at all. If everybody is dependent on the same water system, sharing the flow, then it makes sense to accept a single, unifying structure to society. Opting out, or bucking the system, will mean you don't have any water. Wittfogel said hydraulic societies were despotic by design.

This claim is still debated. Most in the field recognise the impact of Wittfogel's idea, but putting a finger on the exact cause of political development is near impossible. Dr Robert Fernea asks 'does the water determine the shape of government, or does government determine the shape of water?'[40] Tony Wilkinson of Durham University adds to this puzzle by pointing out that the expansion of water schemes in the Sassanian and Roman empires, vital to their strength and authority, must have occurred at a speed far beyond what any bureaucracy could manage. It may be that water schemes were semi-private operations, a kind of benign cooperation between state and local community.

Wittfogel was suggesting that the dam in an arid world was by nature anti-democratic. If you accept the idea, then does that mean democracy is a wet world concept? Perhaps the political tradition of liberty and individualism of the north is not a result of enlightenment, but simply because politics has to be more tolerant and accommodating when citizens are not bound to the state for their water and food? Perhaps the political model of Europe and North America, currently the dominant one by virtue of globalisation, is actually as ill-suited to universal application as Wittfogel found Marxism to be?

Dr Vernon Scarborough, one of the world's leading authorities on the connection between water and history, says that whatever the link between water and ideology, it is an unambiguous sign of power: 'If people are seen to flaunt it (water), it indicates the ability to control things, in the way the poor can't. It's a power game'[41].

No doubt dams and autocrats have always got along. The world's dictators, fascists and Marxists have all loved large dam projects. Wittfogel cited the early civilisations, but 20th century demagogues behaved in a similar way. Much as Mesopotamians, Egyptians and Romans would all have been in awe of the large ditch or dam, so the people of today have been wooed with promises of large waterworks. The political cry is that these would provide power, food and long-term security. Top amongst the various schemes proposed and built were dams. They would come to represent something much greater than a good day out for the city folk. The dam would represent the promise of paradise tomorrow, through unlimited water for irrigation and hydro-electricity, and an end to floods.

When the Prime Minster of India, Rajiv Ghandi, commented on an extensive programme of dam construction, he said 'the drama of harnessing a major river may be more exciting than the prosaic task of getting a steady trickle of water to a parched hectare'. Jawaharlal Nehru called dams 'the new temples of India, where I worship'.

Mao, Chinese leader after the revolution of 1948, had his own huge programme of damming and irrigating. This was the means by which the People's Republic of China would rise to its historically justified position as a great nation again, according to Mao.

When the British withdrew from Egypt, they cancelled their plans to build a super dam at Aswan. The scheme had been criticised on a number of counts, not least that it would disrupt the magical spread of silt upon which Egypt's agricultural miracle was based. However, Nasser, the leader of the newly liberated country, needed a big gesture to show the people they were wise to follow him, that he had a vision for the future, and that poverty could be defeated. So he promised a dam.

He said the country would enjoy 'everlasting prosperity' on its completion. Nasser's rhetoric was inspiring to the Egypt of 1950, but also oddly reminiscent of the old religious texts. Nasser had echoed, in concrete and massive international debt, the monuments of the Pharaohs. He would root the words in the ideological and pragmatic politics of the 20th century, but was selling a dream that was as old as civilisation itself. The co-option of water control into modern politics appeals to the strand of utopian thought that may stem from the promise of paradise in the early civilisations. Like many schemes that promise, in effect, to 'get-rich-quick', dams disappointed. Stalin is famed for building a grand water scheme, promising great wealth but delivering massive poverty. When the waters to the Aral Sea were diverted for cotton farmers, the ecosystem of the region collapsed as the water to the huge landlocked sea dried up.

Wittfogel however stumbled on his own paradox. Hydraulic states may facilitate authoritarian politics, but that doesn't mean the dam in a democracy is necessarily any more honest. Dictators may have loved their concrete edifices, but the west was also partial to great water schemes, and for the exactly the same mix of ambition and corruption.

One of the very few examples of a politically principled water scheme, with no reported corruption, was the 'Hydro' of Scotland. Thousands of Highland Scots fighting against Nazi Germany came from homes lit by paraffin lamps and without any electrical power. A large network of dams, pipes and hydro-electric power stations were built after the Second World War in a bid to bring the living standards of the Highlands roughly into the same century as those of England and America. It was an example of post-war social democratic planning of a kind that would soon become unfashionable.

Elsewhere in the world, dam building mania was driven by a different kind of politics. An emerging global capitalist ideology, perhaps driven by good intentions to give everyone a better quality of life, and perhaps by a desire to make money, advocated dams as the key to development. A few delivered on the promise, but most put nations into debt, and gave little return. For the beginning of this story, we need to go back to the Indus.

The magnificent Indus rushes down from the Himalayas and Afghan plain to the desert of Sindh and the Indian Ocean. When the British left their huge colony of India in 1947, the territory was divided. You might imagine the land with the bulk of the Indus valley would end up being named 'India'. This didn't happen. The land that is home to most of the Indus is called Pakistan. Head of State Mohammad Ali Jinnah invented the name. It means 'pure nation'. To a Muslim, purity and the river were inseparable, so it may have seemed like a clever way of thinking. That didn't stop Jinnah getting highly irritated when Delhi decided to stick with 'India'. How could the land of the Ganges adopt a name derived from the Indus?

Worse was to come for Pakistan. After the Second World War, western leaders had fretted about the shape of the world. Driven by a genuine desire to make things better, and to avoid the financial catastrophe of the 1930s, they invented several institutions that were meant to help the economic development of all. One of these was the World Bank. Pakistan's modern history was greatly influenced by the decision to follow a World Bank plan.

Actually it wasn't so much the Bank's plan as David Lilienthal's. He arrived in the country as another white man with big ideas for the Indus. Europeans had puzzled over how to convert the river into money for a long time. The East India Company achieved this trick by building irrigation schemes off the main flow, which the British Empire expanded into the largest irrigated patch of land in the world before the Second World War. Mr Lilienthal reckoned the land under irrigation could get bigger still. Not only that, but Pakistan could have hydro energy. As its fields were tilled for lucrative crops, its cities would burn with electric light – just like America.

The World Bank embraced the idea and sold it to the Pakistani government. The Bank would lend the money for construction. The new nation would get a ticket to the First World in the form of a huge irrigation and hydro-electric scheme. This story doesn't have quite the happy ending planned. Unfortunately it took the world several decades to realise Lilienthal's plan exacted a huge environmental cost, and a considerable financial one too.

We shouldn't be too kind to him. He was not only an adviser to the Bank, but also a partner in an international investment firm, Lazards. There was outrage when after the second Iraq War the task of rebuilding the country was privatised to Halliburton, and this American firm guzzled up billions of dollars. Lilienthal did something similar. The contract for the water work fell to Lazards. The money was raised by the World Bank and lent to Pakistan – to be paid with interest. The only people who became rich out of this without enduring any risk or consequences were the employees and shareholders in Lazards. Pakistan was locked in a cycle of financial debt and environmental decay it is unlikely to ever break.

The World Bank supported over 100 dam schemes across the planet in the next few decades. These, like the constructions promised by dictators and Marxists, were to bring a bright new world. What they certainly brought were debt to the host country, and a dubious social legacy. The World Bank didn't operate alone. Western companies were successful in persuading states that a dam would be the answer to their development prayers. American Aluminium told the Upper Volta that a dam would be great for making the nation rich. However by the time the government came to sign the contract, they were sold a pup. The poor people of the Upper Volta footed the bill for a dam that has never produced one fifth of the promised power. It gets worse – though rich in bauxite, the mineral needed for the production of aluminium, the contract they signed obliged the scheme to process bauxite shipped in from America.

This chapter could stretch on like a wall of concrete, composed of great chunks of damning evidence against dams and the people who promoted them. They took the profits, and when the metaphoric flood came, in the form of debt, hunger or environmental destruction, they were high and dry elsewhere. The age of dam-building ended in the 1970s, but the legacy is being paid to this day.

In the strange tale of how water and politics collide, the movement to resist dams became one of huge significance. There are famous cases of community resistance, such as the Narmada dam protest in India that involved prize-winning novelist Arundhati Roy, but the one that deserves most attention is the campaign in Tasmania in the 1970s to stop the Australian government building a dam. The protesters organised as a modern, media-savvy campaign, and explained to a global audience the damage this dam could cause.

The victory that emerged out of Tasmania is of particular importance. Let us go back to those Germans on a Sunday outing at the dam. They were innocent participants in an emerging movement. As German identity began to solidify in the late 19th and early 20th centuries, the river and the forest became not just poetic retreats and a pleasant abstraction, but solid ideas. These were concepts that spoke of the soul of Germany, and as such political parties referred to the land and water much in the way modern American presidential candidates might casually reference liberty and the dollar. The idea of 'greenness', once the preserve of pagans, then of poets, became a political currency. At this point, 'green' meant home, and something better than the daily grind.

It is interesting to watch how this concept grows in German politics. Particularly, how the Nazi party encouraged a developed sense of 'greenness' – racial purity had an echo in the idea of ecological purity. Critics have unfairly used this to smear the modern green movement, as if bottle recyclers were close cousins of Hitler. That is no more true than to suggest American president F. D. Roosevelt was a buddy of Stalin's, because both liked building dams.

While post-war Germany was a hotbed of what would become the modern green movement, it didn't have the first 'Green Party'. That occurred in Tasmania, as a result of the dam protest.

To fully understand the potency of large dam and river control projects, and how they intersect with national identity, politics and the advance of human society, one needs to turn to the richest society in history. The USA is the biggest experiment in water control the world has yet seen, and a clear example of how peak water came about.

CHAPTER THIRTEEN

America

'DUELLING BANJOS' is a famous film sound, as familiar as the metallic screech of *Psycho*'s shower scene or the shark's theme in *Jaws*. One banjo plays a simple tune, and another matches this, the two trading notes until they hit a riotous melody. Unusually, we see as the soundtrack is played – a middle-aged urban professional cheerfully accompanies a pale-skinned country boy. John Boorman's 1970s classic film *Deliverance* is anything but happy – the scene sets up a strange lull before the unfolding horror of a canoe trip that goes fatally wrong.

Four urban professionals plan to navigate a mountain river before a dam floods the wild valley. They are out to find their inner manliness, to return to something more essential and free-spirited than modern civilisation. They want to cut loose on white water before it is tamed into a placid reservoir.

The weekend adventurers are city-soft. They talk about feelings and worry about propriety. Only the Burt Reynolds character really wants to go on the adventure; he yearns for the toughness of the wild. As a group, they represent the ambiguity of the civilised, torn between the ease of urban living and the desire to be free. In contrast, the natives of this north Californian wilderness are inbred, hardy and 'uncivilised'.

On re-watching this film, I found it didn't have the obvious chase storyline I remembered. It starts out as a linear tale, of the men going down the river, but a shocking incident jolts the story into more confusing territory, as if the plot becomes stuck in a whirling eddy of water, unable to move forward. Like the police at the end of the journey, the viewer knows bad things have happened, but is unsure of the details. *Deliverance* is a neat tale about how the control of water promises civilisation, but won't stop the stupidity of men. As such, it tells the tale of the American West.

The myth of the west is of survival against nature. It pits a solitary figure against the wilderness and shows how strength and godliness delivers comfort and liberty, often in the form of a log cabin and an open sky. When immigrants were urged to 'Go West' it was to the promise of land and freedom – a new paradise of nature serving man's needs. Like all myths, it existed for a reason – it 'sold' the idea of a hostile place to desperate people as a chance to begin again, and get closer to God, or at least to wealth. Once started, it's a hard story to kick – the lone man triumphing over danger is an ancient tale, loved by all. But it is fantasy.

The conquering of the American West, and thus the creation of an empire which stretched from coast to coast, was a triumph for state control, for bureaucracy, for big bucks. A person settling west of the Mississippi wasn't the embodiment of liberty and a celebration of nature, but the personification of big government, and the brutal taming of nature to conform to man's needs. The west, and by extension America, is the world's largest example of water control, with all the silt that is carried in that historical flow.

If Central and South America were stories abruptly halted and started anew by the European invaders, then North America was virtually wiped clean by the incomers. Immigrants going west literally drove over the traces of previous civilisations. The first settlers at Jamestown came to a continent of developed societies and ancient history. Thanksgiving celebrates the good fortune that the pilgrims lived close to generous indigenous people, who fed and watered the disease-ridden and shambolic Europeans. Two centuries later, the welcome had been forgotten.

It suited 19th century America to think of the west as virgin territory. In doing so, they were behaving just like the Dutch towards their sea marshes, the French and Prussians towards the erratic Rhine and the British towards their myriad rivers – nature was a problem which needed to be tamed, to be put to man's service. The natural was in some way bad, whether it was land that wasn't growing food or rivers that didn't carry cargo. Nature was a canvas on which unworthy aborigines had scribbled childish marks – it was the job of civilisation to wipe the canvas clean and begin again. What had gone before was by definition worthless, because it was uncivilised. The idea of paradise doesn't work if you think you have slaughtered the first tenants, unless they are discounted as non-humans.

The first advance on American soil was driven by the fashions of London. Beaver hats and coats were prized items, and this drove hunters into the interior. Europeans traded guns and drink for furs, in what was a lucrative market for native and outsider alike. A price was paid by the environment. Beavers build dams, which slow down water running off the land. Where water lingers, the soil becomes nourished. For every lady be-hatted in Europe, a stretch of land in America was being diminished. So began a process where the needs of man eroded the fertility of the land. This paradox continues – in search of what the historian Simon Schama refers to as 'American plenty', nature's capacity to provide was harmed.

The priests beat the hunters to the Pacific coast. Spanish missionaries took the Bible and irrigation with them as they founded colonies in New Mexico and California by the end of the 16th century, but it would take believers of a different sect to show how water control could be a miracle that made the desert bloom. The historian Donald Worster refers to the Mormons, who conducted their own exodus to a promised land by leaving the east and settling in Utah, as the 'Lord's Beavers'.

The Mormons were white, northern Europeans *par excellence*. Driven by a religion that embraced hard work and suffering, they settled in Salt Lake in Utah in an explicit bid to build a new paradise. This promised land may have been described as one of simplicity, but it wanted to replicate the level of civilisation found in the old world. As Donald Worster says:

> The stark desert must be subdued, but not at the price of civilised life and living. Somehow they must hold on to the social, economic and spiritual possessions on the conquered desert as well as they had in humid regions.

In 1847 the first turf was broken, and the Mormons set to work irrigating the desert. They were staggeringly successful. With only rudimentary tools, but a strong work ethic, and rigid church organisation, they had over 16,000 acres producing a surplus of food by 1850. By 1890, the acreage under irrigation had risen to over 260,000, and crop yields were up a thousand-fold.

In the story of civilisation, the Mormons fell squarely into the ancient model. The schemes were owned by the church, which took a tithe off the farmers. What was built on Salt Lake could as easily have been dug in Mesopotamia, in terms of the equipment used and the socio-political structure overseeing the work.

However the founding fathers of the United States had specifically determined in the Constitution that church and state should be separate. Washington didn't like the theocratic society growing up on the empire's western border. As with all such schemes, had the Mormons been less successful then maybe the Federal Government wouldn't have bothered. However, it appeared to legislators in the east that the sect in Utah had cracked the problem of the westward expansion, and that was the job of government, not God. They intervened.

The Mormon church reacted by initially decentralising the irrigation system. By de-coupling the water from local government, they hoped to preserve the church's control over irrigation. This was foiled when the Supreme Court declared the move illegal. By 1880 was it possible to operate a privately owned irrigation scheme in Utah with no connection to the church. It seemed then that secular government, that is, the American ideal, was victorious over the old world practice and superstition. However, if the whole of the western USA was to be colonised with ditches and dams, then another powerful, rich organisation would have to oversee the project. The private sector could never cope with the scale of the enterprise. The consequence was that the federal government was substituted for the church. What would be dressed-up as heroic individualism was in fact the largest water scheme by central government the world had seen.

English teacher Katherine Lee Bates boarded the three o'clock train for Chicago in June 1893. She was leaving Boston for a summer school in Colorado. Her journey would inspire a poem. As the steel wheels clickety clacked first to the World's Fair in the Windy City, and then on west, Bates fell in love with the glory of America. Its beauty, wealth and enterprise seduced her to a fit of breathless lyrics.

The verse she created would, after a couple of revisions, become 'America the Beautiful'. The song is favoured by many over the official American national anthem, 'The Star Spangled Banner', and remains very popular. Despite celebrating the new world, its success owes much to the old one. The tune came over with European settlers, and the images invoked are as ancient as civilisation.

O beautiful for spacious skies,
For amber waves of grain,
For purple mountain majesties
Above the fruited plain!

It opens on the glory of wheat and barley fields, and the fruited plain – no difficulty here for an ancient Mesopotamian to immediately understand that modern America was the bee's knees. We then learn, in the second verse, that this paradise was hacked out from hostile terrain, and built by political principle:

> A thoroughfare for freedom beat
> Across the wilderness!

Yet this is not a miserable old world civilisation we are witnessing. It has some new, divine quality:

> Thine alabaster cities gleam
> Undimmed by human tears!

The poem met with immediate success, was soon set to music, and has remained an essential part of American popular identity. Like all good bits of nationalism, it is a benign fantasy imposed on a much grubbier truth. A bit like Bates's journey from Europeanised Boston to the west by train, we must travel through time from the early 19th century, when dreams of an American empire were taking root on the eastern seaboard, to the booming success of the west a hundred years later.

The American empire suffered from the civilisation bug – it wanted to copy and improve upon previous empires. Its written constitution was a deliberate attempt to better the mash of statutes which made up the British system of government, and which boasted about being the home of democracy. So America set out to be more democratic. But having successfully fought the British in the Wars of Independence, they wanted to stamp their new nation with the seal of the old. If emulating London was out, then the citizens of Washington and Philadelphia looked to Paris, and to history. The French Revolution seemed to vindicate the path America had chosen – of liberty and rights under law. But neither revolutionary movement was satisfied with just being new. Both French and American thinkers looked to root their modernity in the ancient dignity of Athens and Rome.

So politicians poured over history books, and littered their new world with references to classical civilisation. They built neo-classical houses, admired Rome and, naturally, they adopted water control as a technique of advancement. They set the army to the task of conquering the unruly water. In 1802 the Army Corps of Engineers was formed. Just as the armies of Germany would tame the Rhine, just as the Roman army had built the aqueducts, the new American army, having beaten the British Empire, would set about making a new empire on this untamed land.

The challenge was the Mississippi. A perfect river, in that it sprawled and flooded and drifted lazily through millions of acres, in nature's scheme, but an affront to the geometry of engineers.

The army picked up what early settlers had begun, in building levees to keep the water in its channel, straightening the path of the water, and dredging the bed. The effect was to turn a marsh into a motorway of commerce. If the initial triumph of America had been the founding 13 states, and the richness of New England, the enterprise received a massive secondary boost by the taming of the Mississippi. Trade boomed. Agricultural yield was up. Success on the river seemed to prove that going west was profitable.

More than that, it showed that America could match the feats of previous civilisations. Roughly a hundred years after the river became the concern of the Corps of Engineers, the first major dam was built at Keokuk, in Iowa. On a water route littered with human intervention, this one particularly inspired a journalist at the *Iowa Magazine*, who wrote that this was 'the most colossal engineering feat ever attempted, not only rivalling, but actually surpassing, the ancient pyramids and the sphinx on the Nile'[42].

As the Mississippi backed up for 65 miles behind the Keokuk dam, so history backed up on this 'colossal engineering feat', the current of civilisations stilling against the rock, the justification of the past flooding the new land. Like the Romans, they had matched the feats of Egypt. The massive importance of the Nile was running through American civilisation just as it had run for millennia. However the challenge of the Mississippi, where there is a lot of water, is very different to that further west, where there is a comparatively little. If America was to match ancient Egypt, it had to find a way of irrigating the desert and the plains.

The west was colonised by opportunists – people on the lookout for a quick buck. Gold diggers, using great water cannons to dislodge soil and run it down hill to sluices, were the obvious ones. While the gold rush was a crucial factor in the colonising of the west, the real 'gold' of the west would be water. In the unusually snowy winter of 1891, an opportunity settled deep and crisp and even onto the lap of a speculator called Charles Rockwood. As the snow melted, and the Colorado River swelled with the thaw, it flooded the Colorado Desert. The soil bloomed, an explosion of green utterly incongruous for this dry region. This inspired Rockwood: if he could build a permanent irrigation scheme in the valley, farming would be possible, and land that was dirt-cheap would become worth a lot of greenbacks. So he bought up land, and started campaigning for a canal.

Land speculation was a huge racket. Eastern merchants were paying Native Americans low sums, and selling on high to their fellow citizens moving west. Some settlers would strike it lucky, either getting fertile territory, or hitting oil. Most would have been as well to burn their money than buy the deeds. At least Rockwood recognised that water was the magic ingredient, the alchemical substance that could turn dreams of a new world into reality.

The turf for the canal was cut in 1900, and completed two years later, leading to 27,000 hectares of irrigated land. To the first farmers there, it must have seemed like a great achievement. History shows that it was a landmark – what emerged from these ditches was a huge irrigated area, Imperial Valley, which is still home to vegetable growers and a $1bn per annum industry. Rightly, history hasn't changed its mind about the ethics of these water speculators.

Here, noble principles, like freedom, counted for nothing. This was plundering on a grand scale. The idea was to find assets, and sell them on at great profit. For Rockwood and his associate George Chaffey, the politics were clear. Water was an asset, and it governed all. Chaffey said, on the success of the Alamo Canal, that 'water is king. Here is its kingdom'.

The problem was that Rockwood and the Imperial Valley were the exception to the rule. Private water speculation was a failure. When left to market forces, the land was dry. To understand why, you need to know about the unique American approach to water rights. What are known as Riparian rights govern rivers in Europe, and while recognising major landowners, they assume that all legitimate claimants have an equal status in the eyes of the law to a river course. In America, it's first come, first served. The original settler on a riverbank has first call on the water, the second settler has second call, and so on. If they don't want the water, they can sell on their right. This may reward courageous immigrants who go west, but it proves an awful nightmare when allocating water.

The first big effect of this unique approach to water occurs in the late 19th century. The Desert Land Act of 1877 allowed settlers 640 acres, on the proviso the land was irrigated. This didn't encourage brave settlers, but rash investment. Companies would buy up water rights, channel the water to dams, and then sell on the stored water to others. The flaw in this scheme was that it's very expensive to control water, but hard to get a good price for it. So all the dollars that went into dams and canals never came back in profits. The companies went bust.

With private enterprise having drawn a blank, people turned to the federal government. When President Theodore Roosevelt gave his inaugural address in Washington in December 1901, he specifically pledged that a new era of water control would begin: 'great storage works are necessary to equalise the flow of streams and to save the flood waters. Their construction has been conclusively shown to be an undertaking too vast for private effort'. With these words America showed it was not so new after all. It would rely on central governance and funds, as every civilisation had done, to build its water management schemes.

The Reclamation Act of 1902 made the president's promise fact. It put into statute what had become a common cry – that irrigating the west was a national mission, a process of taking civilisation to the wild. The Act was supported by the creation of the Reclamation Service, which would later become the Federal Bureau of Reclamation. This body was tasked with acting like an army, but armed with diggers and pipes.

The Service took up the banner of not just water control, but of matching previous civilisations. The completed Roosevelt Dam of 1911 was the tallest rock masonry dam in the world. Irrigation and empire building and civilisation were the same thing to the New World. The tamed water tamed the land and brought the stuff of the old world too. The process of reclamation (from nature, God, or chaos?) was a national task, funded by central government. It was, as such, no different to the schemes of the British Empire, or indeed Mesopotamia. As water control moved west, so did order, and with order came the opportunity to make money.

Where Rockwood had success, and many others failed, was in taming the Colorado. Over 2,000 km of river, pushed on by melt water from the Rockies, it surges down to Mexico and the sea. It was the jewel in the crown, as alluring as the Indus or the Rhine or the Nile. It was, in the eyes of the civilising forces, not a river but an opportunity. In 1912, Californian engineer Joseph Lippincott said, 'We have in the Colorado an American Nile awaiting regulation.'[43]

400,000 hectares in California were irrigated by the river in the 1890s. If the government could improve upon this, then the state would be transformed into a larder for the nation. Oranges, dates and other fruits – the very stuff of paradise – could be grown in quantity. Here was a sense of mission, adopted by politicians and journalists. To be pro-irrigation was to be pro-American. As the journalist William E. Smythe put it:

> (irrigation) not only makes it possible for a civilisation to rise and flourish in the midst of desolate wastes; it shapes and colours that civilisation after its own peculiar design. It is not merely the lifeblood of the field, but the source of institutions.

That explains some of the virulent racism expressed against Mexicans by the schemers and engineers – water control was also about demonstrating superiority over the neighbouring 'barbarians'. The 'source of the institutions' was water control, and such institutions were democratic and civilised, therefore better than the perceived shambles of governance and thinking south of the border.

In 1921 Commerce Secretary Edward Hoover thought of damming the Colorado; the mighty flow could be used to irrigate the desert for agriculture, and turn turbines, giving the new communities electric light and power. To do this, he had to win over the support of the seven states that border the river. Representatives were invited to a lodge in New Mexico in January 1922 to hammer out a deal. They would come up with the Colorado River Compact, which allocated a share of the water to each state, and some to Mexico.

This allowed for the Hoover Dam to be built, along with many others, and for irrigation that would feed countless millions. It also meant drinking water could be supplied to cities such as Phoenix, Tucson, Las Vegas and Los Angeles. The water fed an economic boom that would sustain the whole nation; the south west of the USA is home to 11 of the fastest growing cities in the USA. This suggests the Compact was a great thing. However, there was a problem.

The figures used to calculate the annual volume of water in the river were drawn from an unusually wet period in the century. Nobody in the lodge that January could have known they were basing projections of flow on exceptionally high numbers. However, they must have known that the division was a lie, as they worked on the principle that every last drop was up for grabs. Had anyone cared to think about what a river is – run-off water flowing down to the sea – they'd have realised that if every gallon were accounted for, there would be none left to make it to the ocean. And that is what happened.

The Colorado periodically fails to make it to the coast. This has been the case for decades. This means the salts it should be washing into the sea end up on the land. It also means the last user in the chain, Mexico, is lucky to get a fraction of what it is meant to receive. So the Compact was destined to fail from the outset.

We shall see in later chapters how putting such a strain on a river has a number of other negative consequences, but at this point it is enough to know that the drive for the civilising properties of controlled water put an intolerable strain on the river. It is as if everyone in Los Angeles tried to fill his or her car gas tanks from the same pump, at the same time.

Of course Hoover was successful. Construction of his dam was taken up by the Depression-era President Franklin D. Roosevelt as part of the grand project, the New Deal, which was Washington's way of spending itself out of a financial crisis. The memorial to those who died in the construction of the world's first super dam shows a muscular man standing in front of a wheatsheaf – an image oddly similar to Soviet style art of the same period – and the message is 'they made the desert bloom'. Civilisation's journey was complete. The same motive, to make agriculture succeed where land seemed unsuitable, and the way this challenge galvanised people into a common purpose, had flowed from the Euphrates to the Colorado, from 3,000 BC to 20th century AD. The men in hard hats with power-drills who made the Hoover Dam were in the same civilising trade as the slaves of Mesopotamia.

The left-wing folk singer Woody Guthrie understood this great enterprise as a triumph for the people; they had collaborated through government to invest the wealth of the state in projects that provided jobs and economic certainty. The largest of America's superdams is the Grand Coulee in Washington State. It blocks the Columbia River. For Guthrie, this was a perfect expression of American civilisation, and state intervention in the economy. In 1939 he wrote 'The Grand Coulee Dam'.

> Well, the world has seven wonders that the trav'llers always tell,
> Some gardens and some towers, I guess you know them well,
> But now the greatest wonder is in Uncle Sam's fair land,
> It's the big Columbia River and the big Grand Coulee Dam.

She heads up the Canadian Rockies where the rippling waters glide,
Comes a-roaring down the canyon to meet the salty tide,
Of the wide Pacific Ocean where the sun sets in the West
And the big Grand Coulee country in the land I love the best.

In the misty crystal glitter of that wild and windward spray,
Men have fought the pounding waters and met a watery grave,
Well, she tore their boats to splinters but she gave men dreams to dream
Of the day the Coulee Dam would cross that wild and wasted stream.

Uncle Sam took up the challenge in the year of 'thirty-three,
For the farmer and the factory and all of you and me,
He said, 'Roll along, Columbia, you can ramble to the sea,
But river, while you're rambling, you can do some work for me.

Now in Washington and Oregon you can hear the factories hum,
Making chrome and making manganese and light aluminum,
And there roars the flying fortress now to fight for Uncle Sam,
Spawned upon the King Columbia by the big Grand Coulee Dam.

The American West became, and still is, the most dammed area on the planet. This was simultaneously a triumph for the ordinary man and for big government. The west was the place of artificial dreams from Hollywood, and real ones. Aviation, media and then computers were the big industries, all of which contributed to the process whereby American civilisation became the standard the rest of the globe envied. We saw on the big screen, on the TV channel, a glamorous place, alive with money and technology. It appeared like an ideal world – paradise was the city of angels. And it was all built on great concrete-lined rivers, which made the city possible.

Los Angeles can only function so long as the water comes. Its size and population have far exceeded anything the local water supply can sustain. Which is why the relay of civilisation must come to an end. Los Angeles and the other invented cities of the west became the standard for 20th century urban development. The peaks of downtown LA skyscrapers, and the sprawl of residential plots linked by interminable rivers of concrete and tarmac roads, are what lie behind the extravagance of Dubai. But these cities depend on the great trick at the heart of civilisation, the ability to get water to follow the whims of man. The trouble is that the water is running out.

From the United States of America, it is a short mental leap to the United Arab Emirates. Powered by rapid exploitation of natural resources, determined to emulate the civilisations of the past, confident in man's ability to overcome nature, Dubai is the child of LA, and the newborn in civilisation's family tree. It will be the last of the line. The sound you hear isn't the churchy comfort of Bates's innocent poem, or the happy melody of Guthrie's idealistic folk tune, but the unnerving twang of a banjo, signalling disaster.

III

PEAK WATER

DUBAI IS DECIMATED. Its high towers and glassy wealth shattered after days of war. Most of the people left a long time ago, retreating into the desert, or sailing for elsewhere. The death toll is unknown, but may run into hundreds of thousands. The great monument to 21st century civilisation lies in ruins, shattered into the sand like so many before. Not long before the world had fought over oil, but now water is the prize.

This is perhaps how people imagine water wars. North Africa, the Middle East or India and Pakistan all have critical water supply issues, and longstanding enmities. Also, we are used to thinking of these regions as violent, the place of wars, but it could be different...

Chuck McKenzie stared at the devastation. What had once been the easy place of summer picnics and tourism was now flattened by the great US/Canadian war. Toronto was a drift of twisted steel and broken concrete, the arable fields churned to mud. He knew the same was true of Chicago. And all for what? he wondered, as the rain fell on his pale skin.

A peace was being hammered out over the border in New York. Diplomats scurried around the UN building and offered quotes to the waiting media as a treaty was drawn up. America's thirst exhausted the old treaties and co-operation. Politicians with explosive words convinced tax payers in the USA that Canada was ripping them off, was turning them into slaves, was making the 49 contiguous states a colony of the north. Eventually Washington found it had talked itself into war, and so two of the most wealthy and developed nations, peaceful partners for 200 years, fought.

Could this be our future, our much-predicted water war? Will civilisation stutter to a halt, fighting over the last, vital commodity?

The idea of a water war has become commonplace. It may happen like the scenarios above, but I suspect the world has to face up to a more horrific future. Not one of war as we understand it in 20th century terms, but a state of ongoing global trauma as people witness civilisation decay when the water runs out. How we respond to that catastrophe will be the mark of the human race. Almost certainly it will mean the end of civilisation as we currently know it.

People used to say the war after next would be fought with sticks and stones. This was when a nuclear conflict seemed terrifyingly close, in the decades running from the 1950s to the 1980s. So devastating would that onslaught be that afterwards, the only technology available would be loose rocks and fallen branches.

With the Cold War over, and the threat from mass nuclear deployment apparently gone, we have switched our fears to a water war. What both threats show is that we fear our capacity to self-destruct. In the 20th century we imagined we would do this deliberately, through the use of military technology. In the 21st century we imagine it will be accidental, a by-product of civilisation exhausting the environment.

Yet the hardest thing to imagine is the end of civilisation. It seems a permanent thing. The idea of living without order, or a steady supply of food to the city shops, seems fantastic. If we can picture it, we tend to think it's like a car accident – something that happens to other people. In fact, civilisation has a habit of wrecking itself on the highway of history. The smashes are frequent, violent and utterly destructive.

Civilisations collapse for many reasons. Occasionally the ability to stop decline is beyond human intervention. However, often we are directly responsible for the end; we not only invite apocalypse, we make it welcome.

CHAPTER FOURTEEN

Collapse

IN 1519 EUROPE LANDED in America, again. Columbus had arrived in 1492, while seeking a route to India. Europeans had been coming to north America before that; the Vikings made it, to cite just one example. So it's unhelpful to think of the Spanish discovering the continent – but the time of their arrival is important, as it signals the death of an entire experiment in civilisation.

On 11 September 2001 the Twin Towers of the World Trade Centre in New York were attacked. Both fell to the ground, killing nearly 4,000 people. Few in the world can be ignorant of this, or of the consequences. Now imagine if bulldozers removed the rubble and levelled the ground, and within decades everyone had forgotten the fallen towers, and the dead people, and all the work and creativeness that occurred inside. Roughly, that's what happened in 1519. Euro–Asian civilisation decimated American civilisation, wiped clean the evidence, and everyone forgot. Over the next 400 years mankind behaved as if Europeans were the first to step foot on the landmass. Only in the 20th century was serious attention given to the civilisations that went before.

The Spanish soldiers arrived in what is now Mexico. The most powerful nation on the planet came to the New World with the aim of conquering it and making it part of Spain's empire. European culture and self-image had moved away from recognising the importance of water and on to other attributes of power, such as military might, godly justification and money. Three hundred Spanish troops and the self-confidence of their leader, Hernán Cortés, would defeat an ancient and mighty civilisation that was still in thrall to its water.

The Spanish entered the great metropolis of Tenochtitlan which dwarfed in size Europe's greatest city, Paris. They walked along one of three causeways, which linked the watery city to the shore of the lake. Boats full of food and flowers would have bumped along close to them. Ahead were pyramids and temples and the throb of a city. But quite unlike any European city, this was clean, with covered sewers and a force of over a thousand men charged with sweeping the streets.

Cortés appears unmoved by the scale of achievement of the strongest civilisation in the entire American continent. The leader of the Aztec people was Moctezuma, more familiar to us by the corrupted version of his name, Montezuma. Having showered hospitality on Cortés, Moctezuma and his Spanish counterpart exchanged ceremonial speeches.

Each address reveals the core values of their own culture. Moctezuma speaks to Cortés as if the Spaniard were a god, a messiah-like figure returned from some earlier time, who had a fundamental and supreme claim on the Aztec people.

> You have graciously come on earth, you have approached your water, your high place of Mexico, you have come down to your mat, your throne...[44]

As Moctezuma verbally prostrates himself before the European, he reveals the most precious element on the Aztec empire: 'you have approached your water'. The Aztec empire had built a city bigger than any in Europe, but existing in an arid part of the world. This civilisation had no incentive to move beyond the bounty of that water. The ability to provide crops, hence a secure food supply and in turn stable society was the achievement of the Aztecs'. They had built the magnificent Tenochtitlan as a monument to this, a sign of their power over neighbouring, and less water-savvy, enemies.

For Moctezuma, water control was everything, and as such divine. There was no greater commodity for the gods, no more powerful birthright, than water. It is important to understand that the records of this meeting are not direct pieces of reportage. They were recalled later. Further, the actual speeches had to be translated into the respective languages, and this may have diluted some precise meanings. Had the Spaniards been students of the Mexican tongue, Nahuatl, they could not have missed the importance of water to the Aztec culture. Their word for city is a combination of *a-tl*, meaning water, and *tepe-tl*, meaning mountain. The city, and its monumental pyramids, are literally 'water mountains'.

For Cortés, water would barely have registered as an emblem of state or divinity. Coming from water-rich Europe, and from a culture long ago influenced by the Moors' ingenious water technology, the provision of regular food via irrigation was a given. In such a society, where water distribution was peasants' work, rulers had developed other justifications for their power. Cortés replies by saying that he has been sent by Don Carlos, emperor of the Spanish empire.

> He (Don Carlos) sent us here to see him (Moctezuma) and ask that they should be Christians, as is our emperor and are we all, and that he and all their vassals would save their souls.

For Cortés, the word for city was *cuidad*, which translates as a fort, a place of civic pride, a defensive stronghold. The Aztecs credited water. The Spanish thought themselves empowered by God and weaponry.

Europe's God was to trump America's water. A combination of disease, naivety and technological inferiority would deliver Mexico into Spain's hands. It is not recorded if Cortés, at any point, reflected on the importance of water to the civilisation he destroyed. It is not impossible that he, like so many before and after him in the history of man's relationship to water, just took the liquid for granted.

Yet Cortés had the nerve to call Moctezuma the 'barbarian', that name which serves to mark out the apparently 'uncivilised':

> Can there be anything more magnificent than this barbarian lord (Moctezuma) should have all the things to be found under the heavens in his domain, fashioned in gold and silver and jewel and feathers... In Spain there is nothing to compare with it.

The Europeans based their argument on fiction – they concocted stories of blood sacrifices to the sun god and told these back home, dismissing the Aztecs as simple heathens neck deep in savagery. For what it's worth, Europe in the 16th century was almost certainly killing more citizens on a regular basis, in the name of justice, than any American civilisation. However, we make a mistake to sink into the grubby details of old lies. What is important is that when Cortés described the Aztecs as 'barbarian' he was dressing the business of colonialism in the guise of progress and improvement.

We can move beyond the dumb lies about blood-thirsty sun worshippers, knowing that to be white propaganda, but we can only tread gently when imagining the reality of these civilisations, because we walk on graves. What is known is that a version of civilisation existed, and within a generation had been destroyed.

The importance of Europe's victory over America was that a civilisation which thought itself more important than water control alone had eliminated an alternative, still very much in tune with the necessity of water. Had the USA become a world power in the 20th century with an American cultural heritage, and not the European one that existed, it may never have created such an unsustainable water model.

However, there's little to be gained from wallowing in what-ifs. Instead, we must accept that civilisations do end. Some because of military defeat, such as Tenochtitlan, but many others because they ruin the water on which they depend.

Modern tourists mob the largest pre-industrial city in the world. A newly built airport at Siem Reap, all natural wood and air-con, ushers wealthy people into the poverty of Cambodia. The international visitors are there to view the remains of Angkor Wat. They will get their digital cameras' worth, as this is one of the world's great sites. Stretching over hundreds of acres are temples of varying size, some caught in the root-grip of jungle trees.

Angkor Wat is a triumph of the Khmer people. Under the rule of a succession of kings, the temples rose from the flat plain about a thousand years ago. The scale of this urban plan is too great for any flashbulb wonder to capture. Even if you follow the tourist ritual of climbing a nearby hill and looking down at sunset, you can only discern the towers of a few of the many buildings.

The Khmer people had done what all civilisations do – they had learnt to control water. This patch of south east Asian is naturally wet, and it benefits from the mighty Mekong River, which does something most peculiar – it changes direction at times of spate. The flow gets so high that tributaries cease to feed the main artery, but end up flowing backwards as the volume of water drives the wrong way up the bed.

There is no shortage of water here, just a question of how to control and regulate the water that comes. The Khmer built reservoirs and canals, eliminating variations in the flow and creating a steady supply that irrigated fields, which grew crops, which fed an ever-expanding population, which served an elite. The social, economic and political model of Angkor Wat was broadly the same as we find in many pre-industrial civilisations.

So why didn't these buildings spawn an intellectual and social revival of the scale that was beginning in Europe, in the shadow of the great cathedrals and castles? The gasps of the modern visitors are for a 'forgotten' or 'lost' civilisation – why?

The Khmer lost control of their water system. Not through negligence, but ignorance. The canals ran from the river system to the reservoirs. There would have been other channels that then took the water from the storage pools to the fields. However, the main canals needed to observe a basic rule of water technology – they had to be higher at the source than at the outlet. Like the Roman aqueducts, or the channel of Nineveh, or the British canal system in India – like all water systems, the water has to run downhill.

This must have worked very well at first, and for some time. However the latest theory on why the civilisation of Angkor Wat collapsed is that the canals go too deep. The sheer volume of water flowing through eroded the bed, making the cut ever deeper. The result was that the canals didn't simply run downhill – they ran lower than the reservoirs. Where once the kings had apparently controlled the stuff of life, they failed. That was a thousand years ago and the Khmer people have yet to attain a glory anything like it since.

Civilisation is self-destructive. What it masters, it destroys. While it is a persuasive theory that the failure of the canal system did for Angkor Wat, it is not the only one. There are many reasons why civilisations collapse, and frequently more than one factor contributes to the death[45].

Jericho, the first city, was blessed with a famous spring that still burbles with clear water, but power has never returned to the place. Today, it is a pile of rubble in a suburb – like an empty lot awaiting a bulldozer. That's what happens to man's greatest dreams.

Mesopotamia turned to dust because the climate changed. Around 2200 BC, things got drier. The effect was a drought in the north of the territory. This caused what is perhaps the first historic incidence of economic refugees – the Amorites fled south in search of water. Ur was appalled at the sight of this 'flood' of people, and called the human flotsam 'barbarians'. Despite attempts to build a wall to keep them out, the Amorites settled. Now, the Euphrates, which kept Ur wet, flows north to south, so the drought which had driven the people down also drained the river of its flow. A weakened Euphrates couldn't sustain the needs of Ur at normal times, let alone with an expanded population. Extra crops and food would usually be bought from the north, but there was none to trade. The effect? Mesopotamia withered. The land to the south of modern Baghdad would never recover. Ur is still buried beneath the sand.

As civilisation developed in complexity, so the factors behind collapse become more elaborate. Greece didn't fall because of drought, but war. Rome collapsed by overextending its economic reach and political mismanagement. Though interestingly, it was the 'barbarians' who dealt the last military blows to the once great empire. Proof of their victory came when they destroyed the aqueducts into the city in the fourth century AD.

Humanity could afford to lose Rome, or the other early civilisations. Though many liked to think they were lords of the world, none were truly global. If Rome withered, then the business of civilisation would continue in the Americas, or be revived in northern Europe hundreds of years later. This has now changed. We have one model of civilisation, adopted across the world. The manners and ways of the developed west have been adopted elsewhere. Global trade has linked all people and nations. TV and the internet have shown what's on offer, like a digital salesman promising prizes for all. Many are excluded from the wealth of the world, but all aspire to similar riches.

There are many reasons why we will fail. What follows are some of the watery ones.

CHAPTER FIFTEEN

Luxury

IT RISES OUT OF THE DESERT, its towers and tackiness made soft in the heat shimmer. This isn't Dubai, but a recent precursor, an earlier experiment in making water travel hundreds of miles to reach the city. This is Las Vegas. A perfect example of how civilisation, in its lust for luxury and fun, has relegated water needs to an afterthought.

Water is luxury. The Romans revelled in their playful cascades, literally soaking wet in wealth. Their baths were also steamy with sexual shenanigans and gossip – water is decadent. It's no wonder that the casinos sitting in the Nevada desert reference the classical past. The great fountains outside Bellagio Hotel, or the Luxor Hotel in the shape of a pyramid, or the Venetian Hotel, with its mock canal, or the casino called Caesar's. The mobsters who built this flash, bizarre gambling monument wanted to dip into the same strain of excess, the same fabulous wealth that previous civilisations enjoyed from their water and power. But this was built where there is no water.

The age of aqua-luxury, of hydro-excess may be coming to an end. Not that anybody predicted this when the foundations of the casinos were being laid 50 years ago. Since the north got its hands on civilisation, water has been taken for granted. That is fine in wet places, but it doesn't rain much in Nevada. The people on the strip of Las Vegas are living in a mirage – to co-opt the name of another casino.

In the 1950s an edition of the magazine *Popular Mechanics* speculated on what life would be like in the year 2000. The author has great fun, imaging a place where everything is a lot easier, and better:

> The best way to visualise the new world of 2000 AD is to introduce you to the Dobsons, who live in Tottenville, a hypothetical metropolitan suburb of 100,000 people. There are parks and playgrounds and green open spaces not only around detached houses, but also around apartment houses. The heart of the town is the airport. Surrounding it are business houses, factories and hotels ... Tottenville is as clean as a whistle and quiet ... (he extols

> solar power as the source of energy). Because they sprawl over large surfaces, solar engines are profitable in 2000 only where land is cheap. They are found in deserts that can be made to bloom again.

Effortlessly, perhaps unconsciously, the journalist works through classic themes of water and civilisation. There are 'green open spaces' everywhere, the place is 'clean as a whistle' and the desert can be 'made to bloom again'. It's as if writer Waldemar Kaempffert was deliberately checking off civilisation's perceived prizes, luring his readers, still nursing bruised souls from the Second World War, into a future where everything would be civilised. Imagine reading this in a filthy European city of the 1950s, bomb craters and poverty all around:

> There is no dishwasher. Dishes are thrown away after they have been used once, or rather put into a sink where they are dissolved by superheated water... When Jane Dobson cleans the house she simply turns the hose on everything. Furniture, rugs, draperies, unscratchable floors – all are made of synthetic fabric or waterproof plastic. After the water has run down a drain in the middle of the floor (later concealed by a rug of synthetic fibre) Jane turns on a blast of hot air and dries everything.[46]

The grime of modernity automatically rinsed away by the magic liquid of civilisation, water; this is a lovely idea, slightly crazy yet utterly understandable. The appeal is how simple modern life will be, and the presumption that there will be water enough to wash everything, to start again in the hot blast of a dryer, your suburban dream pristine every morning.

The housing development around Las Vegas really should have internal hoses. They have everything else. These subscribe to what American property people called McMansions, or Plywood Palazzos. The median floor space of an American home was 1,525 sq ft in 1973. It has since grown to 2,248 sq ft, yet there has been no increase in the average size of American families – indeed, as across the western developed world, average occupancy per home is falling. It is therefore hard to understand why houses have not only taken on obese proportions, but also the bathrooms within have multiplied. Where once a single washing facility would have happily served a household, the presumption now is that every bedroom in a McMansion will have its own en-suite bath, toilet and bidet. There will also be other toilets, for the use of guests.[47]

In gated communities, huge structures look out over green lawns, and rolling golf courses and artificial lakes. Never mind that the water for Las Vegas is piped in from miles away – this is 20th century civilisation, where water follows man no matter the lunacy of the destination.

This may be a sure case of ecocide – Las Vegas can only die, and within our lifetimes, because the water supply is running out – but it is currently also good business. These huge houses are just one manifestation of the global property boom of the 1990s and 2000s. As house prices went up, land prices went up, so builders moved to new sites, getting further from natural supplies of water. Not only did they have to stretch the pipes, but then run them into many rooms of each house, as the quality of bathrooms became a key sign for the value of the property. There was a kind of water porn going on, where, strangely, the business of getting rid of human waste became synonymous with high living.

Visit one of the many toilet trade fairs around the world and it appears that the s-bend and the showerhead are the stuff of worship. The bog makers want to create a yearning for better bowls, so they dubbed a random date in the calendar 'World Toilet Day' in 1995. This was meant to be a hook for many positive stories, urging us to get closer to our porcelain. Now 19 November is still 'celebrated' as world toilet day, but by campaigners trying to highlight the desperate sanitation standards for much of the world. 2.5 billion people are estimated to lack decent sewerage, contributing massively to child mortality, disease and economic exclusion.

Flushing a toilet uses about a third of the average household's water. It is therefore no surprise that the 'water closet' was a northern, wet world idea. As a result, it has become a symbol of advancement and civilisation. Cultures once happy to use a more sensible squatting position, and a cup full of water to rinse afterwards, are shifting to the flush toilet. It is just one more drain on the waters of India that western models are replacing the old 'drop' loo.

Of course many in India can't afford a 'drop' toilet. Every day, millions of people pull down their underclothes and defecate in parks, forests, swamps and the out-of-the-way places of their neighbourhoods. This presents a host of public health issues, not least the contamination of water supplies. It is not a solution, but neither are flush toilets. Given that civilisation is something which is copied, aspired to, it is a great pity that Henry Moule wasn't successful in his campaign for a different kind of toilet altogether.

Moule left Cambridge in 1826 to become a vicar in rural England. He seems to have been a determined man, able to upset his parishioners in the pursuit of what he felt to be Christian values. It was when he began the hands-on care of cholera sufferers, in what may have been a related outbreak to the Great Stink in London, he found that buried human waste would decay and disappear rapidly. This was in contrast to the filthy water that seeped around the sick and dying.

Moule found support for his growing suspicion in the Bible, where it says, 'With your equipment you will have a trowel, and when you squat outside, you shall scrape a hole with it and then turn and cover your excrement.'[48] Instinct led him to experiment further, and ultimately patent the world first 'earth closet' in 1860. This was a drop toilet where the user pulled a lever and a quantity of soil covered the mess. Moule had correctly spotted the vanity in the water closet – 'Water is only a vehicle for removing it out of sight and off the premises. It neither absorbs nor effectively deodorises.' What was true in biblical times, and for Moule, is so now. Removing excrement by water is complex, expensive and wasteful. Flushing the stool from your house may appear to be 'clean', but it becomes a problem down pipe. Degrading waste into soil is a better idea.

However, the city is civilisation, and urban life depends on getting water in and out, so water was the reflex solution when cities became stink holes and death traps – wash it all away. That was the triumph of industrial Glasgow, Liverpool and London, and so that was the pattern set for the world. To tell the rich that they should switch to an earth closet would offend their very idea of being civilised. After all, to them, it's a sign of success that they defecate into clean water.

Once relieved, the citizens of Las Vegas might go to the swimming pool, the park or the golf course. While swimming pools have no great impact on water usage, they occur most often in parts of the world where water is more precious for its scarcity. The blue pool, associated with the good life of the USA, is a clear statement of wealth over environment. As the American author John Cheever so elegantly showed in his short story 'The Swimmer', about a man going home by dipping into every suburban pool en route, the pool is an emblem both of middle class achievement, and dissolution. It's a mark of success, and of losing the will to fight.

In the communities around Las Vegas, the love of trimmed grass drinks vast amounts of water. Irrigating a lawn is a wasteful thing. The water must be spread thinly over the surface if it is to keep the grass green. This means it evaporates at a fast rate. Further, grass planted in dry climates is often a matter of centimetres thick, above a sandy or stony surface, which drains the water quickly away, thus increasing the amount you have to sprinkle in the first place.

The city has begun to fight back. Housing developments sold on the promise of green lawns are now, by law, required to let them go brown. There are incentives for planting desert grass instead, and fines for people who waste water. There are limits to this belated water-conciousness. If the community was sincere in wanting to save water in an arid area, it would sell up and close down. The lawns aren't the problem – the city is in the wrong place to start with.

While there are fines for wasting water, there is still a golf course to play on. The amount of water a golf course needs depends on the climate. The home of golf, the Old Course at St Andrews in Scotland, has a sprinkler system. It is only switched on to ensure it still works. Rain provides all the course needs. Of course, this is why golf was a wet, northern invention. Scots would hack a ball through the rough landscape to the green – the smooth surface where the hole was cut. Transplant that invention, which has become a global status symbol, to other climates, and the equivalent of two Olympic swimming pools worth of water, around 4,500 cubic metres, is needed to stop the greens being brown.

The fad for golf in the western USA, the Middle East and Asia is a direct contributor to the stress on water supplies. Courses can prevent poor families from getting an adequate daily supply. Paradise was a place with a wall around it – the barricade is still there between those with water and those without, and arguably it is getting higher and harder to cross with each year.

This isn't the important stuff of civilisation; green lawns, multiple baths and swimming pools are merely the symbols of success, not the essence of it. Much as every emerging civilisation yearns for its water control and economic power, so the individuals have an eye on these vanities as proof of their greatness. That will not make it easy to give up these signs of authority. Yet this must happen. This is the end of luxury, as we know it – those who live in arid parts must learn to forsake the 18th hole.

Similarly mankind must realise that the whims of the 20th century have definitely passed into history. Building a city in a desert, as Las Vegas is, with no regard to water is folly. Somewhere, up river or down, someone's drinking water is being taken so that middle-aged men can loosen their belts a notch as they slump against the fruit machine. When there were not so many people, and fewer demands on water, and the aquifers hadn't already been mined, this was a manageable dream. Now, it is a nightmare.

Spain's water resources are under phenomenal stress, yet a developer has proposed Europe's largest casino should be built on the dry plains of Zaragoza. The promise is a very familiar one. Gran Scala will have buildings in the shape of a pyramid and Roman temples. Julio Barea, of Greenpeace, asks plaintively 'Over here, in one of the driest regions of Spain, where will we get all the water?'[49] The question is whether the authorities have the sense to work out the water implications first, and not be seduced by the promise of easy money. They might take inspiration from a development in the south of Spain. Sociopolis is being built with the wisdom of people who have seen how much of the Spanish coastline has been concreted over (an impermeable layer which causes rain to run into the sea, unable to sink in aquifers) and thousands of houses and apartments built, all demanding more water, all draining water from the farmlands within, destroying the agricultural industry. This new housing site in Valencia doesn't look to modern ways. It is based on the town planning of the Moors. The developer has looked back a thousand years, and is using qanats and small canals. Isn't sustainable water use more civilised than a putting green?.

CHAPTER SIXTEEN

Cities

THE CITY PULLS PEOPLE IN, like a black hole sucking in satellites. It has an irresistible draw, of streets paved with gold and hopes coming true. Yet urban life has always been a mixed blessing. For greater access to wealth, people trade their health and quality of life. When dreams are described by publicity agents, not sociologists, the city is a glorious thing.

During the depression years, when America's wheat growing states were turned to dust by unusually low rainfall, a magical city was conceived to distract people from the awful reality. L. Frank Baum wrote *The Wonderful Wizard of Oz*, in which Dorothy gets picked up from her drought-stricken farm by a tornado and transported over the rainbow, to a land where all she has to do is follow the yellow brick road to Oz, an emerald city where dreams come true.

Baum was writing at the turn of the century, but by dubbing his dream destination the Emerald City he tapped into a truth of civilisation – the ideal place is green, because it is watered. Perhaps it is because this tale is close to many myths, it struck a chord that made the book a best-seller. Set against the events of the century, it is also very sad; in fiction dreams may come true, because of the wealth of water, but in reality, the city has been a nightmare.

From the 1950s on, Mexicans have looked up from the hard toil of the fields and dreamt of going to the emerald towers, which for them was Mexico City. They were doing what poor people all over the world have done, and in ever-greater numbers: moving to town for a better life. However the process of urbanisation has been a disaster for millions.

The city grew out of water control. People were needed to dig and maintain irrigation ditches, so they settled. In return for their labour they were fed by the relatively reliable yield from the irrigated fields. Town and country, urban and rural have always co-existed. In the face of managing our environmental impact in the 21st century, the city is a good 'green' idea; people living close together and being supplied by nearby agriculture. Where civilisation has gone wrong is by putting cities in places with little water, and by allowing the urban mass to swell to dangerous proportions.

Mexico City is founded on the same site as Moctezuma's great city. As the modern urban area grew, the lake shrank, until it was built over. One of the world's megacities, it sits on a dried up body of water. Perhaps there is a metaphor in that; there is certainly a tragedy. What was once the site of a beautiful, advanced city, greater than anything in Europe in the 1500s, is now one of the largest slums.

The city has swollen to proportions way beyond its effectiveness. Like a bus that runs commuters from home to work, with a reasonable load it is a boon. With too many bodies on board, it becomes a tin can packed with sweat.

Part of the fault is ancient. Town planners have always solved the limits of a city's growth by tapping water from far beyond the walls. We have seen how Athens and Rome brought water over long distances, even if this was for vanity rather than necessity. This instinct is basic.

Constantinople (Istanbul) was a city constructed without regard to its water resources. The location was short on wells, but very strong in military and commercial terms, commanding a spot over the river. The Romans put in some canals, and then Byzantine engineers added more. The Ottoman Turks improved the system, collecting water from up to 30 km away. The city grew to be one of the world's largest by 1600 AD, built on those pillars of civilisation: trade, power and water supply.

The Istanbul of the 20th century was New York. Located at the entrance to the Hudson River, Manhattan had been settled as a fort and trading post. To grow as it did, it needed more water, and the tale of this city is partly one of steadily expanding the catchments area for its water. Today, it is working on the latest of many projects to get water to the city; should the water stop, then so does the city.

New York has an advantage. It is in the wet north, and it has political power. By rights, New York should have stopped functioning long ago, but has been able to afford a scheme, equivalent to Colonel Gaddafi tapping the desert in scale and cost, and now a huge tunnel runs for 50 miles upstate towards the great lakes. Assuming climate change doesn't ruin things, this should work for decades.

Mexico City isn't so fortunate. Like a patient in a critical ward, it has tubes running out in every direction, sucking up water to feed the city. However, there simply isn't enough. Unlike New York, the surrounding land isn't naturally sodden. It has drained the aquifer beneath so that the urban mass is literally sinking into the ground. Many live without domestic water services, and more face the taps running dry.

Mexico City's problems arose out of massive growth and quick water solutions. When river, rain and spring supplies were insufficient, the metropolis mined its underground aquifers. It is like a city getting all of its power from an oil well directly beneath. This works fine until the reserve is finished. Then you are in difficulty. So with the water – having pumped up the subterranean lake, Mexico City has nowhere else to go for its water.

A different version of this problem exists for many other cities. In Peru, a land once fertile for civilisations, home to some of the world's earliest and most impressive attempts at civic ordering, the problem is ice. The capital city Lima relies on glaciers in the Andes for its drinking water. Climate change means the glaciers are shrinking and may disappear entirely. Planners reckon it may be all gone within a decade. Millions of people have a few years to figure out new water sources, or they must pack up and leave.

It is unclear if mankind really knows what happens when cities are built. For the benefits, and the undoubted wonder of the city, there is a price. The history of disease begins when people in Mesopotamia switch from the wandering life to the settled one. Suddenly there is a massively increased opportunity for dirty water, for rats to gather, for stored food to go off, or acquire contaminants, and with everyone packed together, any contagion can spread far more easily than among a peripatetic people.

Something else occurs, an alchemy of danger, when humans live so close to one another, and enter the organism of the city. The idea of governance may have emerged from the city, but cities are ungovernable. You can have a mayor and committees, but too much happens in the city at any one time, too many competing needs are met, for any organisation to know what is going on, or what the effects might be. We imagine we are in charge, but this may be a delusion.

Archaeologists digging up Mohenjo-daro found skeletons. The human remains suggest people left in a great hurry, bearing injury, or died a violent death. Something terrible happened to the Indus Valley civilisation, and we don't know what. One of the great places of human ingenuity and organisation was destroyed, quickly, and left no story behind. We have to allow for the possibility that, like the inevitable decline of all empires, so all cities have destruction built in to them, alongside the water pipes and street signs.

I mention this because a trend takes off in the 20th century, and is acutely felt now, of the city itself becoming a force of anti-civilisation. The urban mass, which expands at an exponential rate from 1950 on, ceases to enjoy the luxury of clean water, and clean living. The mega- cities of the developing world gave up, often deliberately, on the Victorian model of providing basic sanitation services to the people. The idea that a pipe would put civilisation on tap either became too much for the city to cope with, or was consciously withheld.

The paradox of post-1950 global demographics is that the process of fleeing to the city has accelerated, but the city will no longer civilise. The fastest expanding urban areas are in the developing world. South and Central America, Asia and Africa are where the city is king. In north Europe (not including Russia) and north-eastern America, city populations have levelled off and begun to shrink. The developed world is no longer quite so enamoured by the city. It jams with traffic, fills with dirt and attracts swarms of people looking for a break. Given our age of environmental awareness we are looking again at the countryside and wondering if life isn't better far from the suburbs.

As it turns out, this sense that the city has outlived its civilising power is borne out in the developing world. The greatest mega-cities also happen to be places where there are water shortages. As the city grew, it sucked up more water. This put greater stress on the rural water supply. No longer is there a neat and necessary relationship between city and countryside, as begun at Ur. Now it is a battle, and the urban acres win every time over the fields.

Where once it was the moral task of city fathers to deliver drinking water and sewer pipes to the slums, in the hope of civilising the dirty hordes, the politics of post 1950s cities has been to deny these amenities to the slums. Now, to come to the city is to turn your back on the village well, and throw your luck into the open sewer of the favela or shantytown. It is, in short, to make an active choice about worsening your life experience, the trade-off being the quest for some hard cash to feed yourself today. The modern savage, if that term were still used, lives in the city.

The dark unknown of modern humanity is the sprawling slum and the street-sleeping itinerants. These are our unwashed, uneducated, threatening hordes, which might smear their dirty existence on the pristine shirts of the rich. The barbarians are no longer beyond the city gates. They are within, and probably built the gates, at less than a dollar a day. Civilisation has inverted itself, and largely because of water.

In 1950 there were 86 cities with a population of over a million. By 2015 there will be around 550. The great population boom the planet has witnessed has occurred largely in poor urban areas. And this has happened at fantastic rates. It took London 110 years to increase seven-fold. From 1800 to 1910, the boom years of the empire, people flowed in to the largest city in the world to make things and earn a living. Compare that with later cities. In the last 60 years, Dhaka, Lagos and Kinshasa have grown 40 times over. The people haven't come for their cut on the profits of a global empire. They have come for the chance of pennies.

Around 1800, the entire population of the world was about 20 million people. Today, there are more people in Mexico City alone. It has grown by a factor of 10 since 1950. Imagine a history where everyone on the planet had gone to live in London 200 years ago, and imagine the state of the city if they had. In your mind is a good picture of modern slums. The Mexican capital has experienced huge growth, but little of it is in the traditional sense of the 'city'. It has been reckoned that 60 per cent of the massive growth is in slum living. In most mega-cities, the shantytown *is* the city, and the old suburbs and centre are the ghetto. The city, that beacon of Mesopotamian strength, of Roman luxury, of British power, of American wealth, is now built of cardboard and disease.

Mike Davis, modern chronicler of the city, puts it thus:

> Instead of cities of light soaring toward heaven, much of the 21st century urban world squats in squalor, surrounded by pollution, excrement, and decay. Indeed, the one billion city-dwellers who inhabit postmodern slums might well look back with envy at the ruins of the sturdy mud homes of Catal Huyuk in Anatolia, erected at the very dawn of city life 9,000 years ago.

At one level we take this for granted. Poor places can't afford nice streets. It's a truism of the modern world. However these slums have been deliberately created. When the British were in India, they knew the value of water control. They used it to build one of the world's largest commodity markets, in cotton. Great stretches of land were converted by the ditch and dam to economically productive acres. The water-politics didn't stop at the fields though. The colonial power also built urban dwellings on the policy of not supplying Indian neighbourhoods with fresh water. Taps were for the governing classes.

This apartheid of water supply is evident in every mega-city. We may assume it is the fault of the slum-dweller. Outside the tax system, they don't deserve or pay for clean water. However, it is more often the active discrimination of town planners who wish to punish the unwashed hordes by guaranteeing they will never be as clean as the city fathers.

When Ernesto Uruchurtu was running for mayor of Mexico City, he knew the basics of politics. Only people with land could vote, so there was nothing to be gained from appealing to the poor. Further, when the aim is to make your city like the chrome-bright places to the north, the last people you need are dark-skinned native Mexicans messing up the place. As the rural peasantry moved into the city, the Mayor had them ostracised. Nothing was more potent, in winning middle class votes, and ruining working class lives, than denying the new suburbs water. He campaigned on the pledge that the water and sewerage wouldn't reach the new arrivals.

To his mind, the native Mexicans were the 'savages'. The irony is that he created a city famed not for its 'civilisation', but for its poverty. His political career ended in 1966, when rival forces recognised the poor people might make eager workers. No longer were the slums for barbarians, but for sweatshop workers. They still didn't get a water supply.

The same process has occurred across the world. The poor come to urban areas, are denied water and other services, so they never have a chance of enjoying the 'magic' of the city, its controlled clean water. In a hundred years the zeal of the Victorian British to 'rescue' the poor from ungodly dirt, and so make the large city a viable entity, has gone into reverse. The slums are uncivilised because they are dirty, and so deserve no water. It's as if the city could be sacrificed, but the prize of civilisation would be hoarded, like a stolen treasure, among those who can afford plumbing.

When Baum whisked Dorothy away to Oz, he pitted her against the Wicked Witch of the West. The cute kid from Kansas has to kill the witch if she is to complete her journey. Summoning all her courage, she does so, destroying her nemesis by throwing a bucket of water over her. As she gets back to the Emerald City, she learns the wizard is a well-meaning fake, and that life comes down to guts and perseverance. She had her ticket home all along, but didn't know it. So does mankind, but it means giving up the trappings of civilisation as we know it. Making the water follow us will not work for much longer for many cities. We are going to have to learn to follow the water.

CHAPTER SEVENTEEN

Irrigation

THE ABORIGINES WILL HAVE the last laugh. The experiment of civilisation has been running in Australia for only 200 years. Before that, for thousands of years, the aboriginal people lived without irrigation, and had no need for writing and politics. They had their own sense of belonging, with an oral history that seems to stretch back into dreams. Before they were banned, tourists would climb to the top of Uluru, dubbed Ayers Rock by the Europeans, and try to imagine themselves a bit closer to the mystery of the aboriginal people. Perhaps the cameras, or the baggage of pseudo religion and self-indulgent philosophising obscured the obvious difference between this land and the west – most of Australia is dry.

The lack of water made the continent a bad place to colonise. It could only sustain a limited number of people. Should more arrive, Australia would resemble an overloaded canoe, wobbling towards certain tragedy in the Pacific. Civilisation as we know it is unsuited to this part of the world. Aboriginal life is much better adapted to the climate realities.

Ecologists and climate change theorists regard Australia as a kind of test-bed for the planet. What happens there will be the fate of many other places. What is happening is that the key relationship of civilisation, the mutually supportive bond between urban and rural, is breaking down.

This was a high-risk colonisation. In very broad terms, wet places are easy to populate, excluding jungles and mountains. As water supplies are plentiful, people go where they want. Ireland is a good example of this. Despite famine and violent political history, people settled on every part of the land. The Irish road map looks like a shattered windscreen – a myriad of tiny lines.

Now look at a road map of Australia. The freeways and minor tracks are clustered on the coast, and mainly the eastern seaboard, which enjoys the best rainfall. Move east and the roads run out. There are no villages and small towns to connect, because there was no water. Unfortunately, there is even less now.

World attention is drawn to this country because of the seven-year drought that has reduced the main river in the southwest to a trickle.

The Murray River basin is home to much of the nation's food and wine production, which is being catastrophically reduced by the lack of water. One might be tempted to see this crisis as a shortage of oaky chardonnay at your local off-license, but the implications are far more serious.

Australian soil requires a lot of fertilisers, which cost money. It also requires extensive irrigation, which costs more money. Thus, there is a premium on the agricultural produce of the nation, which often makes its fruits more expensive than imported food from elsewhere. Australians grow oranges, but they are more expensive than a Brazilian orange shipped to Sydney. Canadian farmers can get bacon on to a Melbourne breakfast plate for less than the native pig industry. The reason Australian wine is so well known is that the profit margin is far in excess of production costs, so it can afford the cost of growing grapes.

The price of urban life in Australia is costly domestic produce, or food shipped in from around the world. As the expert Jared Diamond says, the natural circumstances make it 'expensive to sustain a first world civilisation'. Little wonder the people are clustered in the cities – the countryside cannot support them. Australian cities are like satellites in a sandy universe. Diamond reckons that the nation cannot sustain its current population of 20 million, and would be better suited to a population of eight million. What might happen to the remaining 12 million souls?

If the drought continues, a choice will have to be made between diverting water to inefficient and inappropriate farming, or to the people in the city. Whatever solution they hit upon, it will be a paradigm change in civilisation's formula. A key problem is irrigation.

Had Australian farming relied on naturally irrigated areas for its production, it would have been unable to produce a fraction of its output over the last two centuries. This would have naturally restricted the growth of population, and the attractiveness of the country to immigrants. Instead Australia used civilisation's tool of mass irrigation to grow lots of food, and was able to promise the fruits of paradise.

For the last time in our journey from the beginnings of civilisation to the current time, we need to understand what is meant by 'green'. In the last 50 years, the population of the world has doubled. It has been able to do this because of a quantum leap in food production. This is the phenomenon whereby crop yields in staples such as rice and wheat have been improved four-fold. With four times more food, many more people can be fed. This is known as the Green Revolution. A combination of new seeds, better artificial nutrients and scientific manipulation has resulted in another agricultural revolution. The essential crops of civilisation – wheat, barley, rice and maize – delivered a bounty that expands humankind, just as it did in Sumer. Irrigation plays a vital part in this revolution. Over half the increase in crop production between 1960 and 1980 is credited to the expansion of irrigation. In simple terms, the business of digging a ditch to run water to a dry patch of land is still the most crucial element in feeding the world.

The trouble is that irrigation cannot keep on giving. Nature has several ways of turning off the miracle. When that happens, disaster follows. Drought is the enemy of Australia's ditches. No matter how efficient the sprinkling systems get, they can't work on dust. There are other ways that irrigation destroys the cities it built.

Anyone on a sandy beach in summer, spade in hand and the cries of children flying in the sky like gulls above them, is a proto-irrigationist; you cut a channel and watch the water flow. There is something deeply satisfying about this. I like to imagine that many of the great archaeologists began their fascination with the past on such beaches. Out of the child grew the adult, out of the simple truth of a sandcastle and moat grew the quest for answers from the past.

In the very small world of historians who study water issues, one quotation haunts every written work. The archaeologist Sir Leonard Wooley was a man who had benefited from a childhood in wealthy Britain, with a first class education, who became a student of the ancient world. When we use 'civilised' to denote a wise, compassionate person, Wooley would be a perfect example.

He set out to explain the earliest civilisations. In the 1930s he travelled to southern Mesopotamia, and revealed much of what we know today about the earliest cities in human history. We imagine archaeologists brushing away the dirt, caring only for the hidden treasures below. Wooley was just as concerned by the sand above. As he put it:

> Why, if Ur was an Empire's capital, if Sumer was once a vast granary, has the population dwindled to nothing, the very soil lost its virtue?

On learning that southern Iraq is the home of civilisation, we all feel that puzzlement: how did greatness arise in such a hostile environment? I have attempted to show that the area was wetter and more pleasant thousands of years ago, and how irrigation brought control to the environment and allowed mankind to settle. However, we must address the obvious question – why is the birthplace of civilisation now a windswept desert, 'the population dwindled to nothing, the very soil lost its virtue?'

The answer lies in irrigation. The very control that gave us cities had a fault, an environmental time bomb that would reduce the glory of Ur to dust. This destructive force is a simple one – salt.

All naturally occurring water contains some level of salt. Indeed, there is a multi billion-dollar industry built around our thirst for mineral water – the minerals are salts and the rich glug it down in the mistaken belief it is better for them than tap water. This salt gets into the water from the ground – rain and surface water dissolve the earth's natural minerals. Most of the time, this is a benign process that needn't trouble us. Only in rare instances, such as naturally occurring arsenic that dissolves into some of Bangladesh's water supply, is this process a threat.

So, salty water runs down from the mountains, over the land and into the sea. As this has been taking place for millions of years, the sea is very salty. We all know not to drink seawater, and that it'll be no good for keeping our crops well fed. What mankind didn't realise is that irrigation has the effect of stopping the minerals from running off to the ocean. Water settles on the irrigated land and evaporates, leaving the salt. The natural washing away doesn't occur, so minerals build up to dangerous levels.

This seems like a small issue, as if it could be readily fixed, but salt on the land is enough to bring empires and civilisations crumbling to nothing. It is what killed off Ur. It is the answer to Sir Leonard Wooley's question. Our romantic notion of civilisations is about love stories and bloody battles – a human drama with saintly heroes and black-hatted villains. Actually, human history is determined by less colourful factors. And in this case, the baddy is white.

When man settled to farming, he chose a limited number of plants to cultivate. Naturally, he went for the easy ones – the plants that gave the biggest return for the smallest effort, and ones already suited to the climate. Mankind found grass crops the easiest to cultivate. This is the foundation of the Neolithic revolution, when farming began. From the wild grasses that grew in the Fertile Crescent, the forerunners of wheat and barley proved the most reliable. Our ability to periodically improve the production rate of these two grasses, alongside rice and other staples, allows for the modern world – we have fed ourselves by demanding ever more from a handful of plants.

Wheat is a more fragile crop than barley because it is susceptible to minerals. The records of Ur show a steady decline in wheat yields. From starting out as the main crop, it lost ground to barley. Archaeological studies suggest wheat and barley were equally important at Girsu in 3,500 BC. A thousand years later five times more barley is being grown. In 2,500 BC, wheat was only 15 per cent of the total grain crop. 'Plains choked with salt' is how the bureaucrats recorded it at Girsu. By 2,000 BC the wheat yield in Ur is zero, and accounts for only 10 per cent of the grain from Girsu.

Sumerians responded in what appeared to be the intelligent way. Understanding that salt was collecting on the soil, they figured that if they brought in more water, it would dilute the salt, or wash it off the ground. To this end, canals were lengthened and fallow periods reduced. All the activity seemed to have the effect of reducing the salt rot, and yields of barley increased. This led to greater wealth and a larger population. For a period in the streets of Ur, there must have been that lovely feeling of having defeated a problem by human ingenuity; the warm glow of thinking a once-troubled future was now secure. History would show this to be hubris. The problem hadn't ended; in fact the new methods of farming were sweating the soil of its value, and building up more salt in the system.

Increasingly the irrigation flow didn't solve the problem, it only drowned out the noise of the ticking bomb that would devastate the world's first civilisation. In 2,000 BC, records describe how the earth had 'turned white'. The salt was now so thick on the land, it may have formed a crust. With no fertile ground, there was no food, and with no food, the social and political structure burnt in the sun. Sumer was over.

Crucially, it wasn't the salt alone that killed off Ur. The society could easily have upped sticks and moved a wee bit, or built a new system of irrigation nearby. They didn't, and Professor Peltenburg says

> they were living in, potentially, paradise, but they didn't get it together to change. Short term issues over-rode long term planning – that's what we call collapse.[50]

Today, half of Iraq's irrigated land is saline. Of course, in our global civilisation, that doesn't stop Baghdad having, if politics allow, the same quality of life and social order as Sir Leonard Wooley's descendants in England. What it does mean is that Iraq can never attain first world standards of civilisation on its own resources alone – it has too large a population for its food production base.

Self-sufficiency in food may seem like a false standard – all nations are interdependent in the modern world. However, the process that saw fields go from swaying barley to a white crust can occur everywhere that farming relies on irrigation. The lesson of Iraq is not that it may never be a mighty empire again, or even a stable developed nation. The lesson of Iraq is that it shows what might happen to the rest of the world.

Farming consumes 70 per cent of the world's controlled fresh water. Two thirds of all the water pumped up from aquifers goes to irrigation. Irrigated land accounts for 270 million hectares, and grows a third of the world's food. It uses a disproportionately high amount of water – though we grow twice as much food as a generation ago, we need three times more water to produce it. This is how a researcher put it in a United Nations report on water:

> Vast quantities of water are needed merely to supply, say, 20 mm to a field of one hectare: 20 mm on one hectare is 200 cubic metres of water, and a tanker can carry only a few cubic metres. A quantity of 200 cubic metres of water would last a family of five several years. Each individual needs an intake of no more than about 3 litres per day in addition to water for dishwashing, hygiene, and laundry. In the developing countries, 25 litres per person per day, i.e. 125 litres per day for a family of five, is usually enough. Thus, one cubic metre is enough for eight days, and 200 cubic metres, the amount we need to irrigate one hectare of land, would last the family of five more than four years! Irrigation uses very large amounts of water.[51]

Perhaps it is easier to understand that you need 1,000 tonnes of water to grow one tonne of wheat. That is the calculation an irrigation farmer must make each growing season, knowing that 600 tonnes of that expensive, hard-won water will simply go to waste. Sixty per cent of the water used in irrigation is wasted, in that it evaporates or runs off before it has had any effect on the land's fertility.

This is water well spent, we might argue, when it produces food. However, the consequence of salinisation is that irrigation produces food for a certain time, but then stops. The other great areas of salinisation are, unsurprisingly, the other homes of irrigation – Egypt and the Indus valley, or Pakistan. We are, just like Sumer, utterly dependent on irrigation working, and on its land producing ever-higher crop yields. If irrigation fails anywhere in the world, we will all feel it in higher food prices and possible shortages.

Pakistan's population has quadrupled since 1947, and may peak at 250 million by 2025. It has done so through extensive irrigation and the use of crops with much higher yields than before. However Pakistan now has among the least productive land in the world. Vast stretches of the irrigation system left by the British and extended by governments in Islamabad are barren.

This has made Pakistanis irrigation junkies. For every failure, there is a call for a new dam, or new water system that will rectify the problem. After independence in 1947, and helped by the World Bank, Pakistan built a series of dams in the lower Himalayas. The intention was to 'drought-proof' the country. On one level it has been successful. Since the Tarbela dam opened in 1974, Islamabad can point to 90 per cent of the nation's crops growing in irrigated land, and 50 per cent of the country's energy supply coming from hydro-electric power. Further, it can show export figures, and a generous revenue to the national coffers, from cotton export.

However, 11 million tonnes of salt per year are being dumped on the soil, driving farmers to use ever more water, in a bid to wash the minerals off. This was always a losing game. 40,000 hectares a year are lost. Pakistan is now estimated to have forfeited a tenth of its productive land. Yet its population is still growing. Government hydrologists state that the flow of the Indus, as measured in cubic kilometres, is enough for the nation's needs: the river and irrigation system has 180 cubic km per year, of which growing rice needs 70 cubic km, wheat 50 cubic km and cotton 50 cubic km. But no independent source thinks there is anything like 180 cubic km in the system. If there was, then wouldn't the river reach the sea, at least some of the year?

Pakistan plans more dams. Further barricades of concrete will be strung across the valleys of the Himalayas, in a bid to capture more water, and send it onto the same fields that are currently salting up.

Quite apart from relying on glacial melt water in the mountains, which might not be there for much longer, the scheme is desperate.

The effect of irrigation's promise turning salty is that people who made their living from the land are giving up and going to the city. The slums of Pakistan are growing year by year, with the chance of this condemned population being fed diminishing year by year.

This isn't solely a poor world problem. The only inland desalination plant I know of is on the US/Mexican border. It was built in recognition of the impact irrigation has had on America's western rivers. Water is diverted from the Colorado to farming, and then returned to the flow, so many times that by the time it reaches Mexico it is salty to taste.

The global reach of irrigation is, for the first time, shrinking. Water control is in retreat. In part this is because a fifth of the planet's irrigated agriculture suffers from soil salination. Another factor is the great effort and cost involved.

When the Islamic army reached the old splendour of Baghdad in the 10th century AD they must have imagined their time had come – they owned one of the great cities, a prize for every empire that had risen and fallen in the Middle East. However they were lazy, or simply ignorant of the fundamental reason for Baghdad's glory. They neglected the irrigation system. The splendid canal of Nahrwan, once an astonishing feat of human boldness, cutting 50 miles through the desert, was left to flow for itself. Like so many irrigation systems, this resulted in the silt settling to the bottom. If they ever did realise their mistake, it was too late. The canal stopped running, dammed by its own deposits.

The Islamic rulers, wise in so many ways, appear ignorant to history. When Alexander the Great controlled the same territory in the 4th century BC, he had an army of 10,000 working on clearing the canal of silt. This was a Sisyphean task, for no matter how much silt was dug out, more came every day.

The Islamic armies had been able to capture Baghdad in part because of the Nahrwan cut. A contingent of 15,000 slaves was working on keeping the canal silt free, but in the eighth century AD the slaves had had enough of this horrible task. They revolted and for 14 years, between 869 AD and 883 AD, no silt was dug out. That the slaves eventually returned to work would have been a consequence of necessity. As the years went by without a drained canal, less and less water would have made it to the fields, which means crop yields would have been down. Not digging the Narhwan meant, eventually, not eating.

This weakened the city and made it vulnerable to attack. The Islamic empire may have been in power, bursting with eulogies to water control and greening the desert, but Iraq's crop yields hit a steady decline. By the time the Imams were in town, the region was teetering on the brink of environmental catastrophe. Here, further north than where the fields had turned white with salt, irrigation was exacting a different price; it was drying up. So when the Mongol invaders came to challenge the walls of Baghdad in 1258, defeat for Islam was on the cards. The city, and the region, was too hungry to fight.

The point here is that irrigation involves an astonishing degree of labour and energy. If neglected for any time, it will fail. Mankind has created for himself a task that is vital but which requires constant effort. It may be that in the years ahead, this is changed. In the era of climate change, the idea of huge amounts of energy, much of which is wasted, being spent on growing inappropriate crops must surely end.

'Inappropriate crops' seems an odd phrase. I mean crops which are not suited to the climate or soil. In arid areas, people used to grow plants suited to the climate. It is no accident that the Mediterranean was home to olives, dates and vines. All these have evolved to survive in low rainfall areas. Irrigation changed this. It made the dry farmer feel wet – so water-intensive crops were planted.

History shows this to be folly. The agricultural yield of Iraq dips in the 10th century AD. Some of this is to do with the silting up of the irrigation system and salinisation. However, we also have records that show the crops around Baghdad had changed. No longer were the farmers seeing irrigation as a way of ensuring the production of staples. Now, the flow had become an excuse for luxury. They were growing the crops of paradise, the stuff that was promised in the hereafter. Plums, pears, pomegranates and citrus fruits were in abundance. These crops require a great deal more water to produce satisfactory food. The yield dropped in part because the ratio between water volume and agricultural output was consciously changed – more water was needed per crop. Yet everyone knew the water was being channelled from far away, and at great labour. This shows that the effort to acquire water doesn't result in more careful use.

Today, Australia sends wool to China, wheat to Russia and India. It is running down its own water supplies, jeopardising long-term food production, and threatening water supplies to the cities, while selling water-intensive produce to others. It currently grows more food than it needs, but the population is getting larger.

Irrigation seems to act like a siren call, urging mankind onto the rocks of environmental failure. We are incapable of being satisfied with growing basic crops. Irrigation creates its own mirage of luxury and sustainability. Despite the very existence of irrigation meaning the land is not naturally blessed with water, humans frequently choose to grow water-intensive crops between the ditches. Sophisticated irrigation leads to cosmopolitan farming and fancy fruits. Put another way, irrigation leads to fantastically wasteful uses for water.

It is estimated that 20 per cent of all irrigated land is now suffering from salinisation. This may have some impact on figures that show global grain yields rising on average by 2.1 per cent between 1950 and 1990, but only by one per cent in the following decade. Some argue these figures are distorted by including exceptional years of crop failure, and that human ingenuity is still capable of boosting the yield. What we can agree on is that once-fertile land – and huge amounts of it – is no longer active. However we chose to feed ourselves, we have less land to do it from.

If food consumption continues at the current rate (famines and starvation still with us) then we will need to increase the amount of irrigated water by the equivalent of 24 Nile rivers. That is two dozen new Egypts, if you like, if we are to meet future food needs. The formal way of expressing this is to say we need 2,000 cubic kilometres of irrigated water. On the surface this seems impossible – there just isn't the land suitable for such a mammoth task. The solution might lie in making our gallon of irrigation water go a lot further.

The irrigation specialist Sandra Postel puts it thus:

> we need to double water productivity – get twice as much benefit from each litre of water we remove from rivers, lakes, and underground aquifers – if we are to have any hope of fulfilling the water requirements of eight billion people and protecting the natural ecosystems on which economies and life itself depend.

For Sydney, Melbourne and Adelaide, the clock is ticking. At the best of times, water is a precious commodity. The drying up of the Murray

River is seen as a rural/agricultural tragedy. In fact it is a tragedy for civilisation on the continent.

The aborigines have had a hard time in modern Australia, with a higher than average chance of suffering social exclusion. They can sympathise with the slum dwellers of the world. They might even laugh when the cities of Australia begin to fail. These people have been on this huge island for around 50,000 years. The standard view is that they came to a tough region and learnt to cope. An alternative view is emerging.

Scientists think that for a long time man enjoyed a lush, green environment in Australia. The aborigines started off in a place that was close to ideal for human settlement and development. Then something happened – some kind of human intervention that changed the rain pattern. Around 14,000 years ago, according to this model, the land switched from being wet and rich in food, to an arid place. The theory is that a form of climate change took a green and pleasant land and made it a desert. In short, the aborigines messed up long before the Europeans arrived, and long before civilisation tampered with the planet's system. It is possible that the aborigines made a similar mistake to that now occurring in Australia, and have had to live with the consequences for millennia. Perhaps none of us should laugh after all.

CHAPTER EIGHTEEN

Farming

WHEN THE AMERICAN ARMY invaded Iraq, with support from the UK and others, hundreds of thousands of young men from the most advanced society in the world arrived at the home of civilisation. And they were, like all young men, hungry. So a supply chain was created, providing a western diet of burgers to the sands of Mesopotamia.

Feature articles highlighted the burgers, a way of reassuring relatives, friends and countrymen that the boys at war were okay; they were eating meat, so things couldn't be too bad. You could picture them, jarheads in cotton t-shirts, elbows on the table and hands to mouths.

Food aficionados regard the burger, and fast food, as the decline of civilisation: people burrowing into their hands as they walk along, their noses deep in spongy bread and processed meat, with ketchup running down their chins. Food should be savoured, enjoyed slowly, and prepared with more love than a teenage worker at a McDonalds can muster, is the critics' argument.

Leaving aside snobbery, the burger may contribute to the future of civilisation. The production of meat, so loved by the developed west, has a direct bearing on the control of fresh water. While the US soldiers sank their teeth into the patties, and probably felt their hearts lift with the sense of peace and pleasure after the stress of armed duty, they were biting into an issue more important than the fossil fuel over which they were fighting, however strange that sounds.

Of all the water controlled by man, 70 per cent goes on irrigation. Domestic use is a mere fraction of this, and industrial use is a declining share in the developed world and clocks in at around 20 per cent. In short, the world's main water problem is about farming.

As a species, meat has been our luxury item. We are omnivorous as animals, but the effort required to kill a beast is far greater than that needed to collect berries or plants, so most of the time we've relied on a herbivore's diet. You can measure the relationship between effort and diet in the list of animals made extinct by human hunting. We've killed off the easiest prey. The auroch, the ancestor to the modern cow, is long gone, as it was slow and simple to catch. The buffalo is faster and possibly smarter, but hopelessly vulnerable to human hunting methods.

The big switch, in terms of the history of civilisation, from a largely vegetable diet to one that included a lot of meat occurred in medieval Europe. The reason is surprising – plague. The Black Death swept over northern Europe in several waves, cutting the population down like a scythe through grass. This meant there were fewer labourers to till the fields. Labour-intensive foodstuffs like root vegetables and grain were harder to nurture. Keeping beasts, whether cows, pigs or sheep was easier. Once the labour of building an enclosure was done, you needed fewer workers to raise your crop.

A head of cattle presents a number of problems for the world's water resources. This lowing, benign animal is a multifaceted industry on four legs. The meat will be eaten, either as good cuts or the mechanically recovered flesh that is mulched into burgers. The skin can be tanned into leather. The milk sloshes into the many products of the dairy industry. The cow serves us well. Unfortunately, it exacts a hidden but high price. To raise a single beast takes a phenomenal amount of water.

Imagine preparing the food for a soldier in Iraq. The wheat for the bread and burger bun represents 150 litres of water. If you want a milk-shake that'll be 1,000 litres, the burger will set you back 3,000 litres and an ice-cream afterwards will represent about 1,500 litres. If the soldier had to pay for his meal in water, he would need to drive a large tanker to the canteen, just for lunch.[52]

If meat were an occasional luxury, this might be a hydrological price we could easily pay, but as the western model of civilisation and progress is adopted across the planet through globalisation, we face a major problem. Meat consumption in the newly wealthy China and India is doubling.

As the water expert Philip Ball puts it:

> livestock rearing places an additional stress on water consumption, requiring around 60 billion litres each day globally. The meat rich diet of the western world is fantastically wasteful of resources – there is simply not enough land and water to allow all the world to indulge such tastes.

Vast quantities of irrigated land are given over to growing water-intensive crops, which are then fed to cows. Half of the world's wheat, 80 per cent of its maize and most barley, is fed to pigs, chickens and cows. It takes 11,000 litres to grow the feed for the cow. The dying acres of southern Texas and Mexico are draining the last ounce of damp from the Colorado so as to grow alfalfa sprouts for beef cattle. The Rio Grande is a desperate trickle so that American citizens can eat burgers as an everyday meal.

This is not the only problem of beef farming. Apart from the water the cow itself drinks, and the water that goes into its feed, the mass demand for beef has had a lasting impact on the land. Livestock require open grazing, which means trees and other plants are removed. The long-term effect of this is that the land loses its natural defences against soil erosion. Where once a mesh of roots would hold the fertile land together, the flat plains of grass are easily washed away. There are an estimated 3.3 billion grass-eating livestock – that's sheep too – and they are reckoned to have denuded around 20 per cent of pasture through over-grazing[53].

Water that once dallied on the land, depositing mineral riches, now rushes by, clogging rivers with the earth it has gathered up in its flow. This is what happened to the people upriver on the Euphrates, and drove them down to Sumer thousands of years ago. They were following their rich soil as it was washed away. What looks like control of the landscape, by removing trees, can easily lead to degradation. Land which once held water, and therefore the possibility of settled life for people, becomes uninhabitable. Each year, an estimated 10 million hectares of crop land is lost to erosion.

The troops in Iraq probably didn't give a lot of thought to the way a burger affects the world's water supplies. On the surface, they were there to address a phantom threat, and establish democracy. Dig a little deeper, and we know they were there to protect the region's oil reserves for western interests. The important stuff was underground. In that respect, the soldiers might have some understanding of the water crisis. We have pumped most of the water wells dry.

The well is the very first example of man's attempts at water control. In the 8th millennia BC, wells up to 10 metres deep were dug in what is now Cyprus. This is a fascinating mystery. At the time, Cyprus had a wet climate. The well locations are near natural springs. There seems to be no need for the additional water. To make matters stranger, the holes were dug before the discovery of metal, when digging 10 metres down would have been very hard work. Archaeologists guess the wells were an insurance against dry years.

If digging a well was once an act of prudence, of long term planning against the worst, it has become quite the opposite. Civilisation has used underground water in one great splurge, like bankers burning money in the name of growth. In India, for example, the green revolution tripled rice yields from 1960. This allowed for the population of the country to increase at one of the fastest rates of anywhere on earth. There are now over a billion people who need their daily meal in the nation. Post-independence from the British, a large dam building programme was begun, but only partially completed. The water from these dams is significant, but does not explain the massive increase in rice production. That is down to the diesel generator and the pump. Indians have been tapping into their underground water to feed themselves. One expert put the cost of pumps at $12 billion since 1980, thought to be extracting 250 cubic km of water a year. Unfortunately for India, rain replaces only 150 cubic km per year.

Western campaigners became agitated when Coca Cola set up in the south Indian region of Kerala. The drink-maker was accused of stealing water. This was an easy fight – bad global company selling sugary fizz, against nice local people protecting their food. The truth was less easy. The Real Thing was extracting water, but at nothing like the rate of agriculture. Tracts of land in Tamil Nadu and Gujarat are now barren. The water beneath is gone. The farmers are going to the city, to join the slums. Campaigners would be better protesting at the local landowners sucking India dry than Coke.

The United Nations Environment Programme (UNEP) says that groundwater, like that dragged up by the belching, smoky pumps of India, slakes the thirst of two billion people. It is, according to UNEP, diminishing. Which is alarming, when you realise 40 per cent of global agriculture relies on aquifers for growing crops. That we must change is beyond dispute, but how is a question of debate. Some gather up all the gloomy water news and say we are heading to hell, and we'll get there within a decade or so. Others say things are brighter.

Growth rates for rice, maize and wheat have declined. The New York based Worldwatch Institute says this is a portent of doom. Sceptics say not at all, as they point to the selective use of statistics, and to topical factors, such as the current inefficiency of Russian agricultural production. Rather than become drunk on these details, I would prefer to keep sight of the bigger picture. Water in traditional wheat, rice and maize growing areas of the USA, India and Pakistan is running out. It is possible that the wet north could replace production, in part. However, this raises important geo-political questions. Does the arid south want to rely on the wet north for food, and where will everyone get their daily glass of water, if the wells run dry?

CHAPTER NINETEEN

Trade

LET US START WITH A ROSE: a symbol of paradise and something purer than the grubby reality of the everyday. It is packaged, marketed and sold to us like so many commodities in the developed world: immature buds, wrapped in cellophane, getting dirty on garage forecourts or looking sad in the buckets hawked around night spots by immigrant workers looking to make small change in expensive cities. Who thinks of heaven on smelling a bunch of flowers bought at the petrol station?

The west and north inherited the idea of the rose as a symbol of love, royalty and fidelity from the Middle East. Out of the earliest ideas of paradise came the flower, twisting its thorny way into the consciousness of most civilisations. We know that there must be a huge production machinery behind the ready availability of the blooms, but choose not to think about this when declaring our adoration for the girl in the coffee shop.

In fact, the production machinery exists in places like Kenya. Around Lake Naivasha they grow flowers. It started with one farm, but it was so successful at nurturing tight little buds for the west that others followed. It is part of a trend in East African farming. Producers had spotted they could get a lot more money from items for the western market than they could supplying the domestic table. The reach of the western city had gone round the world in search of sustenance.

However, flower growing is very water intensive. The rose growers weren't just another type of farm, but a particularly expensive one. The flowers were sucking the lake dry. The consequence has been that the water level has dropped, the aquifers are dry and the farmers now have to look at alternatives, because there is just not enough water to keep their petals from going brown. For people who backed the rose revolution, they got about 15 years of good production before their investment turned to dust.

Worse happened to the people who weren't flower growers, but were trying to make a living on and around the lake. The water consumption of the roses was so high, it took water from other crops, and other land. Soon people had to abandon the country and go to Nairobi for work. At the lake, a war of the roses had broken out, between cash hungry growers and angry locals. As ever in water wars, no one won. The water has gone.

What happened at Lake Naivasha occurs across the world. Things are grown or made that require large amounts of water. An individual rose might need 10 litres to reach the pathetic little ball of petals that is traded as a flower. In effect, Kenya was exporting water to the west, to the rain lashed streets of Brooklyn, Glasgow and Berlin. If we really valued the flowers, that might just make some sense, but they are disposable tokens in the industrial west. Kenya sells precious water to the wet north, only for it to be trampled underfoot after a failed date.

Global trade is based on the unspoken idea that it is okay to take water from one part of the world and export it to another, in any quantity you like, so long as it is in the form of something else. Unfortunately there are no glass and steel buildings buzzing with shiny traders who are making millions from the trade in water. For the lack of this handy image, connecting the water trade to other global commodities, we imagine there is no market. And it is true that there are no financial instruments like a water bond or water futures, and that prices in water are not bartered on computer screens. However, there is a global trade in water.

To understand this, we need to understand how much water is used in the production of everyday items. To grow one kilogram of rice can take up to 5,000 litres of water. A kilo of wheat will put you back 1,000 litres. A pint of milk could have taken up to 4,000 litres of water. The principle that applies to food production works with all produce. A steel beam, a computer or a fancy dress all require water to be made.

So when I buy a bag of rice from India, costing me around $3, I am also buying up to 5,000 litres of Indian water. Given the cost of transport, marketing and other business expenses, the actual water is being valued at close to zero. We needn't do the maths to realise the water is coming dirt-cheap. Tony Allan of the School of Oriental and African Studies in England dubbed the trade in water as used in the production of food and other goods 'virtual water'. He estimates that the world transfers around 1,000 cubic km of water. Imagine the entire flow of the Amazon not running along its bed, but zipping across the world alongside container trucks and cargo ships.

Virtual water is like the idea of food miles. Food miles were invented as a way of getting western consumers to understand the energy that went into transporting their food to the supermarket shelves of Edinburgh or Boston from the fields of the world. They measure the distance a runner bean, for example, has been flown in order to get to the check-out. If we were to adopt virtual water as a similar scheme, it would show us the volume of water consumed in the production of all items. With the virtual water scheme, you could pick up a bag of sugar and see that it needed 3,000 litres of water to be grown.

Virtual water measures the amount of water used in the production of a single item. It shows that water is traded, but disguised in other commodities. However, it doesn't show us the relative use of water. Take a head of cattle for example. As discussed, huge water resources are required to grow a cow, and sustain a western, meat-based diet. But if the cow is raised in Scotland, so what? There is lots of water. Raising a Scottish cow doesn't deprive anyone else of his or her water supply.

Raise that cow elsewhere in the world, and you may have a major impact on water. Many of the irrigation schemes in south west USA go to fields that grow cattle feed. As we have seen, these irrigation schemes cause horrendous problems – turning much of the land infertile through salinisation, and severely denting Mexico's legitimate claim to the same water, forcing Mexican cows and farmers to go hungry. So, raising an American cow can mean a much bigger water 'price' than a Scottish cow, simply because water resources are under much greater pressure.

Having a single virtual water measure for the amount of water that goes into a cow is helpful, but is only part of the picture. To get the real value of that water, you need to know where it comes from. America is a net exporter of cows. It is also a net exporter of wheat. As a result, it is estimated that America exports around a third of its controlled water, which is the water that has been channelled to farming. So large tracts of American farming land are short of water, and are harming the fertility of the land through irrigation. However, the produce from that land gets sold to other nations.

If I take a tonne of water and use it in the manufacture of something like a laptop computer or steel beam, then I can reckon on producing around 70 times more commercial value from that water than if I use it to grow food. Water used in industry attracts a far higher premium than water used in agriculture. In simple terms, this explains why food is very cheap, historically, but cars still cost a lot. We are prepared to pay a lot for material items, but a little for our food.

It is not just America that commits this self-destructive trade. Australia exports water in the form of sugar and wine. Israel, bound up in the water dance from hell, still exports the precious liquid in its tomatoes, much as southern Spain does in its citrus fruits.

This would be a very good thing if we diverted 70 per cent of our controlled water to industry and only 20 per cent to farming. That would mean 70 per cent of our water was earning a big mark up, and thus making people rich, giving them the money to invest in ever more efficient water schemes. However, it is the other way round. We divert only 20 per cent of our controlled water to industry, and 70 per cent goes to farming. Worse than that, industry in the west is getting ever more efficient at using water – peak water use in industry occurred sometime around 1970. But agriculture is using ever more water.

This situation occurs because we do not 'trade' water. We attach no value to its use in agriculture, the largest single use of water. A tonne of wheat grown in water-stressed Egypt is priced at exactly the same as a tonne grown in the lush fields of Ireland, for example.

That is why I suggest a 'water log'. It takes the idea of virtual water, which is the measure of how much water has gone into a product, but then relates it to the amount of water available in that part of the world. Thus, in illustrative terms only, a Scottish cow might have a rating of one – indicating that all the water it took to grow that beast was readily replenished. But a Texan cow may have a rating of 10, indicating the cow took 10 times more water to grow than Texas could easily spare.

So, any food item should carry a water log label. Consumers could decide if they wanted a vast bowl of oranges at Christmas, grown in southern Spain from stressed irrigation systems, with a rating of say seven, when perhaps they should treat fruit as a luxury rather than a cheap foodstuff left to rot in the bowl.

I don't see the need to attach a water log to all manufactured items, as industry is alert to its water use, and market forces more rigidly govern the cost of production. No one makes a steel beam for the hell of it – they have a buyer, and a good price, in mind. However, it might be helpful if a 'water log' were attached to cotton. Like food, the price of cotton has crashed. If I look in my wardrobe, there are numerous cotton shirts and t-shirts. Cotton clothing is so cheap, western consumers can buy items casually, not worrying about throwing them out if they don't like the shirt after all. Yet cotton sucks up the earth's water in gargantuan draughts.

If consumers begin to discriminate against food and cotton that comes from water-stressed areas, then those nations will earn even less from their produce than before. This would seem to be counter-productive. Subsistence level farming will collapse – millions will go hungry, because their farmland is geared to feeding international markets.

Maybe we have to explore another approach. In simple terms, what if food and cotton were not cheap? Hydrologists and food experts are clear on one matter. Cheap food has led to bad irrigation – there is no profit to be had investing in better water distribution. If we paid producers more, they could invest more in water efficiency. If cotton was worth more, you could grow less and yet earn a bigger profit. The only way we are going to do this is if we give a notional value to water.

As a system, it helps everyone. A Scottish farmer can grow cows while Saudi Arabia can save the $40 billion spent on water-pumps and irrigation schemes for cattle feed. This isn't an easy victory for the wet world. You cannot grow cotton in Scotland. Cows, sheep and barley – yes, but cotton, wine, olives and much more are a dead loss. Water-logging might result in the price of cotton going up, making cotton growers richer. It would signal that the age of cheap food and cloth is over – if you are going to turn over your land to produce items for international trading, make sure they are ones which are suited to water resources.

The idea of water having a cash value is hotly disputed. We resist the stuff of life coming with a price tag. In some sense this is strange. Water has been sold for a very long time. No sooner do you get the emergence of the city than the water-seller appears, hawking the liquid on the street for those who are thirsty, or who want a clean supply for cooking at home.

In modern times, it is only the French who have always assumed water was a utility you paid for like any other. In the 19th century, when French cities were developing their water supply and sewerage systems, it was private companies who were in charge. As a result France is home to some of the largest private water operations in the world.

France is also where bottled water first became big business, the leading brand being Perrier. The owners of the distinctive green raindrop shaped bottle had been in business for a hundred years selling the naturally carbonated water when they decided to expand. America was the target market. The advertisers realised selling water was going to be tricky when it ran free out of the tap, so they invented a sense of status around the product. To have bottled water was a sign of prestige and wealth.

Thus Perrier became a cool thing to have, and earned $110 million for the company in 1989. The next year disaster struck. Traces of a poisonous substance were found in a bottling plant in the USA, and the entire worldwide stock was recalled. Sales collapsed, and the food giant Nestle bought the business. So successful was Nestle at marketing the bottles anew, thus encouraging countless others to enter the market, that the American bottled water trade was worth $4 bn by 1997. Ironically, given the backlash against the product ten years later, on environmental grounds, the boom had been started by the invention of polyethylene terephthalate (PET) plastic bottles, which were claimed to be recyclable. Now, the fashion in top dollar restaurants is to ask for tap water.

The tale of bottled water shows our ambiguous relationship to water pricing. On the one hand the west readily embraced the old idea of water-sellers, so long as the stuff was 'cool', but just as quickly dropped it when fashion changed. While people were paying $4 for a bottle, they were broadly opposed to the idea of water privatisation, and of paying for their domestic supply. Further, while the rich world worried about the differing qualities of various brands of bottled water, the poor world has largely engaged in a fight to resist the World Bank's latest scheme – to privatise water systems.

For all the money that the World Bank has offered towards water privatisation, few schemes have been successful. In broad terms, people don't like the idea of western companies making a buck out of the right to water. The main water companies – Suez, Lyonnaise, Thames Water – have tried to turn the water systems in parts of Africa and South America into profitable exercises. Most have failed. Indeed, so unproductive a policy has it been, that private enterprise is withdrawing. The UN's second world water development report in 2006 claims private water companies are quitting developing countries – 'privatisation is a heavily politicised issue that is creating social and political discontent and sometimes outright violence,' says the report.

This may seem a victory for the people of the south's big cities, but in truth it often means they are left with no water supply. As for the companies, they are concentrating on the wet north. Paradoxically, their profits are to be found in nations that do not struggle for either water or money. There is more profit in persuading rich westerners to upgrade leaky pipes than there is in getting a basic daily requirement to the people of the south. For the north, water privatisation appears more like an exercise in pumping life into neo-liberal capitalism than a necessary protection of infrastructure[54].

Capitalism is interested in water. There are several funds on the New York and London markets that are betting on water developments. The main issue is how the world will trade water in its liquid form. The problems are many. Water is heavy, awkward and costly to move. It must be clean of contaminants, which means careful handling. Pipelines are expensive, and vulnerable to sabotage. Like oil, even if you do pipe your water in from afar, the well will soon run dry.

Currently, bottled water aside, water is traded in two forms: as a part of other products, as discussed above; and in bulk by carrier. Several locations around the Mediterranean welcome huge tankers filled with water, to make up for old and over-used water supply systems. A Norwegian firm is testing vast balloons filled with fresh water that can be towed by ships. Thus it appears a similar system to the trade in oil is emerging.

However, water isn't oil. It is the stuff of life. I find it hard to conceive of a similar distribution system, equivalent to petrol stations and tankers, for water. The energy required in shipping it, the sheer amounts needed, and the expense of distribution are factors. More importantly, if you live in a place where there is only expensive imported water, I suspect you'll move. It's more likely that people will go to the water, not wait for it to come. It's just too important.

Perhaps Bill Phillips could help unravel the conundrum of how to value water. In 1949 he was trying to settle an argument about how the British economy worked. John Maynard Keynes and Dennis Robertson disagreed on how money circulated. As computers were still exotic, Phillips modelled the flow of cash using water. Gluing together fish tanks and hosepipes, he constructed a bizarre machine, resembling some glorious lunacy at work. In fact, Phillips was very bright, and his homemade contraption worked. A central tank filled with water, showing an increase in the money supply. When it overflowed, this represented income, which whooshed down some pipes and split, showing the divide between taxes and savings[55].

This lovely contraption, useful at the time, is kept by the Bank of England, which currently worries about how to keep capitalism afloat. Things might be better if we worried about other things than money. Perhaps our behaviour should change. Civilisation is a model of living which is suited to societies that control water. Those societies tend to mimic each other, and ones that have gone before. Civilisation is, to that extent, a 6,000 year game of copycat. It may be time for the wet world to set a water use example to the arid world, so that the idea of civilisation moves away from huge water consumption. In the process, we may also shift our idea about what wealth means.

We should be the ones who build new houses with composting toilets and reed beds to clean the waste water. We should instigate rainwater collection on a large scale, so that every new building captures some rain for use in the toilet or garden. We should ensure we have low-flush toilets. We should ensure that more food is grown for local consumption. The wet world should grow vital food for the dry world. We should demand of the European Union a food policy that doesn't let naturally irrigated land become idle. We should exert further pressure on industry and farming to be efficient in its water use, and we should punish harder those who pollute.

We should not do all this out of guilt. It would be silly if those in the wet world, where water shortages are less of a problem, were to live as if the water supply was a fragile resource, just because we felt bad about people elsewhere. The reason we change is that we value civilisation. For the first time in the history of civilisation, success should be marked by the careful use of water, not squandering it.

The reality of changing our water use is colossal. It will come up against the fact that water distribution doesn't respect national boundaries. Global trade is based on the sale of commodities, not the protection of resources. It calls for a new kind of civilisation, built on global co-operation. The penalty for not doing this will be widespread social chaos that will claim millions of lives.

As climate change takes grip on the global imagination, a new attitude should emerge. Our fingers must be prised off the idea of progress at all costs. We must accept that the next generation may get a lot less 'stuff' than the last. Perhaps we are entering the age of uninvention, when we are no longer slaves to the latest advance? Perhaps we should take longer to grow flowers that smell of paradise, and more time to enjoy them.

The reality will soon hit home. California's water illusion isn't simply draining on the Colorado. It is massively expensive to run. Twenty per cent of all generated energy in the state is used to move water around. That's a fifth of the output from power stations and natural gas plants. If carbon footprints are to be shrunk, and climate change gases reduced, something has to be done about that vast expense of energy. Peak water isn't just about water, it's about how we live.

A British newspaper wanted to examine the reality of the annual Human Development Report, published by the United Nations. The 2006 version states that 1.1 billion people do not have safe water, and 2.6 billion have inadequate sanitation. A journalist was dispatched to the slum of Kibera, which spreads out from Nairobi, the capital of Kenya. It is swelled by families who once lived on farms at places like Lake Naivasha. Writer Ashley Seager summed up the global dilemma thus: 'The people, constantly ill with diarrhoea, pay more for their contaminated water from standpipes that those in Britain do for their mains water.'[56]

CHAPTER TWENTY

War

THE KING OF GIRSU looked from his city walls at the watery mess below. He was all-powerful in the hierarchy of the irrigated society of Mesopotamia. To him fell the responsibility of overseeing the water, and distributing the harvest. As such, life flowed from his decisions. What he saw in the flooded fields was a direct challenge to his power.

The canal from the Euphrates that supplied Girsu had been sabotaged. The water was no longer under control, but ran unchecked over the crops, drowning them. The man responsible for this act of terrorism was the King of Umma, who wished to gain control over Girsu.

Girsu could have responded by rounding up its slaves and attacking Umma, but instead the labour force was charged with building a new canal, this time from the Tigris. Once completed, Girsu had the better water supply, so was able to grow more crops, and became more powerful than Umma.

This episode may constitute the first war in human history. It was over water. By war, I mean something more organised than the tribal fighting of pre-civilised societies. War is a very civilised thing – it occurs when two settled, organised societies aim to acquire the wealth of the other. It requires the structure of civilisation to be waged, and is driven by an essence of civilisation, the acquisition of greater resources.

The first water war of the modern era occurred in 1947, when India closed supplies to Pakistan on the partition of the two countries. It has dragged on to the present day. The second was in 1967. In six days, the well-equipped forces of Israel took land from Jordan and seized the Golan Heights from Syria. This territory is still occupied, and subject to international dispute, along with Israel's treatment of the Palestinians in Gaza.

From Girsu to Gaza, there appears to be a long history of water wars, and contemporary conflicts that make future ones likely. The adage used to be that the war after next would be fought with sticks and stones. This presumed mankind would obliterate civilisation with nuclear weapons, and be reduced to fighting with blunt objects. Internet wisdom now says that the next war will be over water. It might be, but I think we need to understand the particular nature of water disputes before assuming there will be organised carnage over a river.

There will be no war memorial for Ken Proctor. The 66-year-old Australian didn't die at Gallipoli or in Iraq. He was killed on his front lawn, in the suburb of Sylvania in Sydney. He died when a neighbour mistakenly thought that Proctor was breaking drought restrictions by watering his garden. Angry words turned to fists and that was that. It turned out that no by-laws were being broken. Proctor died from a mistake, and for the strange need to keep a lawn green in an arid environment.

Disputes that get out of hand occur every day, over any number of trivial things. However it is important to realise that when water becomes so valuable people are prepared to fight for it, it will not be the kind of organised battle that has recently occurred in Iraq for oil. Everyone, from the army general, to the owner of the general store, will be involved. This is why water wars are the conflict of last resort for civilisation and humanity.

There is some intrinsic urgency about the idea of war that suggests something must be done, soon, and with extreme force. No amount of evidence against the need for war may be enough – the primal call of violence shouts louder than facts. However, in this instance, those yelling about the likelihood of water wars, and the next global conflict being about 'blue gold', are likely to find events not quite so predictable.

There are places in the world where water is contested between two or more nations. There have been, and always will be, fights over water. Blood will flow wherever water goes. But war is a tricky thing if the prize is water. Say there is a river which two nations need. Nation A is downstream of Nation B. So Nation B blocks off the river, diverts it into irrigation schemes, and leaves Nation A with only a trickle. Firecrackers and flags for Nation B, you might think. However, within days, Nation A will have become a lawless state – without sufficient water, the bonds of civilisation snap; individual survival is all that counts. Nation B, still recovering from the hangover of its celebration party, will find a wave of people pressing into its territory. Not just a few thousand, who might be repelled by a fence or border patrol, nor a hundred thousand, who may just be deterred by tanks and missiles, but millions. And these people won't be united by their old sense of national identity – they will be prepared to do anything to quench their thirst. Unless Nation B is prepared to countenance slaughter on a colossal scale, the victory it won over water supplies will soon fade into chaotic human tragedy.

Lets say muscular Nation A decides to invade Nation B, to take control of its water. Tanks roll, fighter planes swoop and somehow Nation A is victorious. However, it now has a much bigger problem. If it is to retain control of the water supply, it must stay in Nation B forever. Unlike wars for oil or other commodities, you can't just pump up the water and leave. Nation A would have to occupy Nation B for the long term, and suddenly it must deal with all the inhabitants of the newly acquired land. They will need water too, if they are to be governed. Nation A has merely increased its problem.

War is of questionable use in the control of water. That has yet to stop powerful nations from threatening it, or arming themselves on the basis that a fight for water is imminent.

Egypt is a relatively wealthy nation. That is not to deny that many Egyptians live in poverty, but measured in national terms, the place is doing okay. Certainly it is rich enough to pay for a large military operation. You might be tempted to think that these planes were bought for flying north, when Egypt was enmeshed in dispute with Israel. In fact, the squadrons of shiny metal are primed for sorties to the south. What Egypt fears is Sudan and Ethiopia.

This may seem strange. Compared to the cotton wealthy Nile nation, a hubbub of trade and tourists, its southern neighbours are poor and dishevelled by war and famine. Between them they couldn't begin to afford the jets. However the threat they pose Egypt isn't in military might, but water control.

The source of the Nile was a beguiling mystery to the western world in the 18th and 19th centuries. The historian Simon Schama suggests that north Europeans' search for the source of rivers reflected a key change in thought. Protestant determinism supplanted Arabic and Asian fatalism. History no longer happened, it was driven along a course, like a river. There was a beginning and an end. The search for river origins marked this shift in literal terms – humanity now put great effort into finding the exact spot at which the Nile began. Though this makes for a great Victorian adventure, one can also see how it is ultimately a meaningless quest.

Still, astonishingly brave men set off, native assistants at their side, following the river back. Schama hints that this was like a journey in time, as if the north European explorers were venturing to a source not of the actual river, but the river of life. They were on a mission not only to find where the water began, but where man began. As it happens, getting to the source of the Nile would bring us close to the first Homo Sapiens, but of greater concern to the men on these journeys was a sense that if they found the source, they discovered justifications for the power of their nations. If Britain could manage such a grand feat as being first to the spring, then it gave the empire the veneer of justified superiority. Britain's explorers were unwittingly connecting to the earliest known forms of religion, in which power sprang from water.

For Ethiopia, where the Blue Nile begins, power does spring from water. In the highlands, rain falls in copious amounts. The landscape is like any other well watered one – green and lush, hills rolled smooth by the flow of the water, healthy people living off the land. This part of the nation is well fed and secure, compared to the more well-known desert lands where periodic drought makes people suffer. It's no great leap of the imagination – it's one every other society in a similar position has made – to reckon that if only the benefits of the highlands were extended to all, then the nation would flourish.

The perceived threat to Egypt's water supply led Nasser to build the High Aswan Dam. He did this to deliver the blue promise; controlled water would revolutionise the nation. Of course, the Nile is the site of the longest civilised settlement in human history. The bounty of the flood has meant Egyptians have always enjoyed a good deal from the global water lottery. But Nasser wanted Egypt to meet modern standards of civilisation – American and European standards. He needed to increase the productivity of the nation. The hydro power from the dam would make Egypt 'first world'. It seemed like a straightforward decision to the nation builder.

His blue promise has held true. Egypt has enjoyed massive economic growth, largely a result of the extensive irrigated land producing cotton and wheat. The water supply has been guaranteed by the vast reserve held behind the concrete monument of the High Aswan. But at what cost? Egypt is, by any measure, losing its water war – namely the war to preserve sustainable supplies of the liquid. The evaporation from the High Aswan is catastrophically high – as Nasser was warned at the time. Much more sensible if he had followed the foreign engineers' advice (it came from old colonisers Britain, so perhaps carried the whiff of something rotten) to allow a dam to be built in the hills of Ethiopia, where the cooler climate would mean much reduced evaporation rates. But that wouldn't have made Nasser look strong. Many nations emerged from colonial rule in the 20th century, and most of them wished to acquire the symbols of strength that would prove their legitimacy; hindsight may suggest this was wrong, but hindsight doesn't win elections or respect.

Thus Egypt is addicted to the dam – loss of its supply would cripple the entire economy. Any suggestion, as Ethiopia regularly makes, that the Nile's resources are controlled in a different way, is swiftly rejected. Egypt gets 97 per cent of its water from the Nile, and three quarters of that comes from Ethiopia. An old treaty from colonial days gives the lion's share of the Nile's flow to Egypt, some to Sudan and none to Ethiopia. The fear in Cairo is that one day Ethiopia and Sudan will take matters into their own hands. Those planes and tanks and troops are kept in readiness for a war for water.

The Egyptian army, despite the treaty, must also keep a keen eye on its northern border. The truce with Israel might not hold if the region descends into outright conflict. If there were to be a war, water is a very likely cause. Israel, Jordan, Syria and Palestine are all living on borrowed time. A combination of too many people, growing water-intensive crops and living western lifestyles, means the region has run out of water.

The territory that Israel occupied in 1967 was the River Jordan's catchment area. Rain flows down from the Golan Heights into the Jordan. Israel needed the land to guarantee a secure water supply. It accused Syria of planning to dam and redirect the water from the Golan Heights, a claim denied by Damascus.

If it can happen once, so the reasoning goes, then another war in the region over water is possible. This gains greater likelihood when you consider that the water reserves of the region are under much greater pressure than in the 1960s, and the per capita availability of water is reduced. Add in the fact that Israel, backed by the United States, has a huge military and nuclear arsenal, and you have the stuff of disaster.

Undoubtedly water is a point of disagreement. The Palestinian people are effectively being starved of water by Israel. Israel is also taking water that Jordan legitimately claims. Israel occupies Syrian land, held in the name of water security. Israel's water needs are rising, as it attempts to maintain a first world quality of life to show that it is part of the 'western civilisation'. It thus has high water consumption but in an arid region. That said, it's among the most diligent water conservers – there is, however, only so far water conservation can go.

In 1964 Israel dammed the Sea of Galilee and pumped water out for its national grid. So efficient has this been, that for nearly 20 years no water has flowed from the sea to the River Jordan. In its pursuit of western standards of living, Israel also pumps up ground water from the aquifer that partly sits under Gaza, where the Palestinian people are encamped. This has been too successful; the aquifer is low enough to allow water from the Mediterranean to seep in, thus contaminating the supply. The people in Gaza depend on water tankers from Israel. While Israel is splashing in its last drops, Jordan is already virtually dry.

Israel cannot return the Golan Heights without signing its own death sentence. Without that Syrian land and water, Israel cannot function. Jordan will soon grind to a dusty halt. Gaza will be uninhabitable. War looks very likely. 'There is no water in the Middle East. Therefore, understandings must be reached. If not, it can turn into a war or forceful confrontation,' says Uri Saguy, head of Merokot, the Israeli national water company. However, no amount of tanks will bring the water back. For that, all parties would need to change their agricultural habits, thus changing their economies and foreign trade positions, and changing their quality of life. Perhaps war is the easier option?

There is another water war that could begin a thousand miles to the east, between Pakistan and India. The two nations emerged in 1947. The colony of India that the British governed until that year stretched from the western Pakistani border to the eastern Bangladeshi border. On independence, the territory split into three political entities: India, Pakistan and Bangladesh, which was then under Pakistani control. This was motivated by religious difference, with Muslim politicians declaring the need for a Muslim state. This prompted a great shift of people over the subcontinent to Pakistan, the promised Islamic land. India chose not to become a Hindu state, in contrast to the plans in Islamabad, but to stay a secular democracy as envisaged during the long campaign for independence from Britain. The result has been 60 years of bitter enmity between the two nations. Bangladesh sought independence from Pakistan in 1971, and thus opted out of this saga.

Both nations have large armies, and both station a significant proportion of their forces on the shared border. Both have some nuclear capacity. The dispute began soon after Independence, when India began to dam a tributary of the Indus. Both sides have since used water as the medium of attack. At issue between the two has been the water that flows off the Himalayan mountain range and into the rivers that feed the Indus, which is, as we have seen, the lifeblood of Pakistan. While both dispute the rightful ownership of Kashmir, the northern Indian region that Pakistan claims, it is the water that really matters.

Take the Chenab River – it starts in India, and flows into Pakistan and the Indus. In 1960 the World Bank brokered a deal which gave control of the Chenab, and two other Indus tributaries, to Pakistan. Another three river rights were given to India. However India has started to build a dam on the Chenab, the Balglihar Barrage, in clear breach of the treaty. India says it needs the dam for hydro power, and the flow downriver will not be affected – however, it also refuses to go to arbitration.

To compound this, in 2004 the Indian Punjab unilaterally annulled all water treaties, thus removing its duty to leave some water for neighbours in Pakistan. The greatest danger is for the 150 million people who live in Pakistan, as the Indus supply is dropping at an alarming rate of seven per cent per annum. The Strategic Foresight Group of Mumbai predicts a crisis by 2010 for Pakistan's agriculture. Little wonder an ambitious Pakistani soldier wrote a paper in 1990 on the Indus Valley as a cause of future of conflict. He was Pervez Musharraf, who would come to power by coup in 1999. He has since left office, but the threat remains. Pakistan, for reasons of a desperate populace and a broken economy, may find war with India a convenient answer.

However, that is not the full picture. Within Pakistan, the Punjab region is verbally attacked by politicians and farmers from Sindh for stealing water. In India, the middle regions are increasingly looking with envy to the wetter north for water. We think of water wars being between nations because that is what we are used to, and because there is a political structure that gives voice to national grievances in military terms. However, civil war is just as likely. Without the Indus, there is no reason or method for Pakistan to exist. Like its northern neighbour Afghanistan, it may find a return to tribal identities more suited to the new water reality.

Civil war was the result in 2003 when Sudan pinned its hopes on a grand new irrigation scheme. The planned bounty never came, and the country was left in considerable financial debt. Worse though was the social effect. People began to move to the wetter areas, prompting a war of particular ruthlessness. The rest of civilisation was apparently unable to stop the killing. Again, like so many conflicts, there were other factors at play, of identity and politics, but the failed control of water was a key factor.

So there could be international war, civil war or class war. This last category may seem quaintly outdated. I mean there is a clear sense of rich and poor in most societies and the poor are the ones without water control. Whether it is street riots in Bolivia, protesting at water privatisation, or the aftermath of Hurricane Katrina in New Orleans, water control and violence are closely connected.

The investment bank Goldman Sachs has dubbed water the 'petroleum for the next century'. The report points to an increase in global demand, which is doubling every 20 years. And so we are encouraged to think of water as oil, a resource that will bring out the usual responses of governments, but water is more important than oil, and it provokes a different reaction.

Professor Aaron Wolfe has assessed the 412 international crises between the end of the First World War, 1918, and 1994. He discovered that seven had water as a factor of any kind, of which four involved any fighting. That is, three of the seven saw not a shot fired. Wolfe concludes, 'As near as we can find, there has never been a single war fought over water'[57]. While you may quibble at that finding, it is worth remembering that from 1950 to 2000 there have been 507 disputes between countries over water. The statistics would seem to show that most disputes don't end up in war.

Eric Lomborg says there have been 3,600 international water treaties in the last 1,200 years, and 149 have been signed in the last 100 years. He argues that the threat of a water war is counter to the facts. People have an interest in co-operating over water like no other resource.

Co-operation on water is very resilient. Throughout the 1965–75 Vietnam War, the Mekong Committee on water operated, negotiating fair water use for all. Israel and Jordan fought for 30 years, and held secret water talks throughout. The truth is that water may be just as effective at disarming man's war instincts as it is at starting the fight.

The word rival comes from the Latin, *rivalis*, which means to share a riverbed. The Permanent Court of Arbitration in The Hague handles water disputes and it says 263 river basins are disputed. Will these rivals go to war? The Water Co-operation Facility of the United Nations thinks not. Put simply, diplomats reckon the costs of a water war are so high peace is better.

Does that mean there will be no war? I think conflict is very likely, but of a scale and kind we have never witnessed before. We stand on the brink of the new, the unknown, and at stake is the very essence of civilisation and life.

CHAPTER TWENTY-ONE

Dubai

I AM WORRIED ABOUT MY LEGS. They are white beneath the water. This is not because of some strange chemical reaction – I am always corpse-pale. My Scottish skin was not meant to lounge in swimming pools in the United Arab Emirates. As the temperature hits 46 degrees, I may burn like phosphorous in the sun's rays.

To me, this a fantastical temperature, almost other-worldly. I could as well be on another planet. Before coming, I half-wondered if normal behaviour would be possible. At such heat, could anybody move? It was like a journey into space.

Dubai is very close to the experience of a moon colony for pale-skinned northerners. Caucasian expatriates live in gleaming air-conditioned buildings, move in chilled cars, eat in ice-cool restaurants serving food from home. They live a sealed-in life, behind air-con locks and dollar-rich entitlement. This is how NASA would design things; the outside is to be avoided for a good quality life.

Dubai has shot up from the ground like a concrete leylandii plant, freakishly big in the blink of an eye. The tallest tower in the world pierces the sky, a canyon of skyscrapers dwarfs a multi-lane high street, there is a man-made archipelago consisting of islands shaped like continents and a housing development which springs from the coast like the fronds of a palm. These are internationally recognised symbols of Dubai's chutzpah, or hubris. Is this civilisation?

A hotel called Atlantis wows customers with an 11 million-litre aquarium. There is a plan to make a year-round ski resort in the mountains to the east of the city, with 'snow' that can withstand 35 degrees centigrade. The editorial of *The Nation* newspaper throbs with self-regard,

> The UAE has consistently defied expectations by breaking into ever-growing numbers of unlikely markets. Most famously, the country ignored the advice of so-called experts to become a regional centre for shipping, airlines and tourism.

It goes on, 'The technical challenges are formidable, but the possibility of being able to schuss in a swim suit is just too wonderful not to welcome.'

2.4 million barrels of crude can earn the UAE anything from $100 million to $300 million, depending on the price of oil. That is in a single day. It has made the country giddy with cash. The UAE is capitalism as a fun park – a whirlwind of fleeting happiness sustained by petrodollars.

As a race, we are driven forward by the sense that things will get better. Our religions say heaven will improve on earth. Our history suggests man moves from a cave to a luxury apartment. The promise of capitalism is that more money or stuff will make us happier. In the film *Blade Runner*, an advertising hoarding flies above the rotten streets of a future Los Angeles, its audio message offering people the chance to 'begin again' on a new planet. We can tell this one has had its day – it rains constantly in the Californian city. As the world's finances collapse, there is a renewed sense that we must begin again, our streets are rotten. Is Dubai the future?

It never rains in Dubai. Part of the city's success is the promise of good weather to pimple-skinned northerners who want to get out of the cold, much as Los Angeles boomed because of its favourable climate. But fleeing to cities in the desert doesn't feel like progress. It is a symbol of mankind's technical ingenuity, but it feels like escape, or running away. There is, after all, nothing else here.

The UAE has marketed itself as a distillation of the rest of the planet. Not only are there islands shaped like countries, but the country has lured in foreign investment on the promise that this is a capsule of everywhere else. No matter your nationality, you'll find something of home here. The strap line of one development, Falcon City, boasts that it is 'The world in a city'. The UAE hasn't sold the world its past. Instead it is marketing a glossy future, much as *Blade Runner* imagined space colonies as a place to 'begin again'.

It makes this promise just as the world is looking a bit sick. The economy here is built on oil. Carbon fuel powered the last century, but we know it must run out sometime, and the damage it causes to the environment makes many wish it would stop flowing tomorrow. 'Black gold' is the fuel behind the globalisation and we must find something to replace it. In order to begin again, we are looking to science and what it can do with 'blue gold' – water.

This seems a strangely circular obsession. When in doubt, we turn to water. The sixth century BC Greek Thales of Miletus wanted to explain the world in a rational way. He suggested, for example, that the earth floated on water, and quakes were caused by huge waves hitting the land. Of more importance was his suggestion that all things were made of one substance. This is the first expression of the atomic idea. Thales didn't say atoms were the building blocks, but water.

Water stayed at the heart of intellectual enquiry. Medieval alchemists were interested in the art of turning water into other substances, though they obviously failed. The magician Paracelsus is thought to be the first person to make hydrogen in an experiment. As the scientific revolution of Europe gained momentum from 1500 AD, water was often the subject of experimentation. Hydrogen was named 'water-maker' by the 18th century scientist Lavoisier while he was testing the qualities of the liquid.

Now we ask water to solve our energy needs. There is great hope in hydrogen power. It produces energy when hydrogen is mixed with oxygen; the waste is pure water. Currently, exhaust pipes drip with condensation as they purr with noxious fumes. It is a nice idea that our cars may motor along, leaving trails of fresh water, like an eco-conscious snail. It appears to be a win-win situation[58].

Hydrogen power seems to slot in to the long, and wonderfully inventive, history of perpetual motion engines and schemes for free energy. None of them work. There is always a price to pay. Hydrogen power may rescue us, but it takes a lot of energy to produce the pure hydrogen needed as fuel. That hasn't stopped the 'water-maker' attracting hype: 'Hydrogen-powered fuel cells promise to solve just about every energy problem on the horizon' says the expert David Stipp[59].

There is an industry body for hydrogen power. Unsurprisingly, it believes it has the answer to the world's power needs: 'Hydrogen is the next logical stage (after oil and gas), because it is renewable, clean, and very efficient' says T. Nejat Veziroglu, President of the International Association for Hydrogen Energy[60].

Something about water suggests it will provide miraculous solutions. It seemed that way when two scientists, Stanley Pons and Martin Fleischman, announced they had cracked cold fusion: the production of energy by fusing atoms together, in a glass of room-temperature water. Briefly the world rejoiced, until it proved impossible to repeat the trick, at which point the men were denounced as charlatans.

What they did prove, in a glorious explosion of enthusiasm and sweet dreams, was that water remained the magical quantity for human imagination. Somehow the whole illusion was both more incredible, but possibly true, because of the water. Our yearning for alchemy, and faith in H_2O, was as strong in the late 20th century as it had been at any time in history.

One of the big attractions in Dubai is a water adventure park – a mix of pools and slides. It performs its own alchemy, convincing people they are in a 'normal' pleasant environment. The truth is that this city is an invention of water engineering, and is a crazy place to build. Like all desert tricks, it has a short life span. The water is already running out. Dubai has the highest water consumption per capita in the world. It is situated in one of the driest parts of the world. You can see how this isn't going to work.

Inland from the luxury coastal cities are sand and oil. Perfect lines of empty tarmac lead out to the Empty Quarter, a huge desert. It is from this area that the Qarasyh came, Mohammed's tribe. Road signs warn of oil rigs crossing the tarmac. Parallel to the road are four strips of newly planted bushes and date trees. They run the full length of the 100 km or so journey. They are linked, like a daisy chain, by black hosepipes at the roots, providing water in the heat.

On the edge of this undulating strangeness is an oasis called Liwa. Something odd is happening here. The oasis, a point where ground water is very close to the surface, has been exhausted, but the pumps are still working. The towns of Hmeed and Mazaira function, and there is even a palace on a hill for one of the Emirate royal families, dusty and empty but for a few soldiers and groundsmen.

At night the sky is perfectly black. In the desert there is no human light to spoil the stars. The magical stretch of space is supposedly the next destination for civilisation. Exploration of the moon and stars was initially driven by super-power ego. Now the machines flung into the darkness are sent to look for useful things. If we have exhausted this planet, then we'll need another one with water.

Opinion has swung between there being planets with rivers and oceans and the view that cold space is water-less. When a map of Mars was mistranslated by the American author Percival Lowell space promised not just water, but civilisation. In 1877 Giovanni Schiaparelli mapped what he had seen through his telescope, and noted linear marks as 'canali', for channel. Lowell read this as 'canal' and, ignoring the mockery of his peers, fixed on the idea that civilised people lived on the Red Planet.

> Girdling their globe and stretching from pole to pole, the Martian canal system not only embraces their whole world, but is an organised entity. Each canal joins another, which in turn connects with a third, and so on over the entire surface of the planet,

he wrote, adding: 'There is a network of irrigation ... certainly we see hints of beings in advance of us'.

Within years, H. G. Wells had concocted a 'war of the worlds' when the creatures of Mars did what all civilised empires do, and went to war. Thus science fiction and plain mistake have influenced our view of water in space. By the 1960s opinion was erring towards an arid universe, but then a sequence of discoveries changed this. In 1969 molecular clouds were discovered, which contained water. The water molecules emitted energy that can be recorded by radio telescope. H_2O wasn't Earth's alone. Life may be a freak occurrence of this planet, but water is common.

The *Mariner* craft observed riverbeds on the moon. Mindful of Lowell's mistake, there was caution before anyone declared lunar amazons at NASA in 1972. The evidence appeared overwhelming, but the punchline would disappoint the optimist. The rivers had flowed, but over three and half billion years ago. Evidence of a shallow ocean led scientists to give it a name, Oceanus Borealis. This had existed about a billion years ago.

In 1996 a lunar mission discovered what appeared to be an ice lake at the southern pole. Two years later another unmanned orbiter encouraged scientists to estimate there was around six billion tonnes of frozen water. Much of this is in the Aitken Basin, on the dark side of the moon. The consequence of never enjoying the sun's direct heat is that the ice is minus 230 degrees centigrade. Our nearest planet has water, but it is locked down at a temperature so low it is of no use to humans.

This appears to be the pattern. The heavens hum with the radio wave frequency of H_2O. The pulse of water can even be found in the blasting heat of the sun. However, only earth is blue. Space is black and ice-white, Earth is colour.

The water that exists on Mars is underground. This is a consequence of being smaller than Earth. As a lump of molten metal and a target for meteorites, it must have had a similar evolution to that of our home, but it was able to cool faster. This meant Mars created a fixed hard crust. Earth has a crust that moves. This means we have an atmosphere, and Mars doesn't. Water is recycled not just from the surface of the earth, but from the centre too. On the Red Planet, it is stuck beneath.

Venus has an atmosphere, but not one conducive to a watery civilisation. The extreme heat of the planet, 500 degrees centigrade, means the water is steadily evaporating beyond its reach. Astrophysicists estimate the planet will be waterless within 200 million years. That may sound like a good shot for a new civilisation, but who could bear the temperature?

If you hang on to the idea that our future rests on another planet, then pick a moon of Jupiter. Europa, Ganymede and Callisto have icy surfaces and salty currents beneath, with a magnetic field too. Perhaps you may 'begin again' there?

We are still spinning hope, and the promise of a better tomorrow, around the control of water. A theme of civilisation is that we travel on, our backs against the stream, placing our faith and fate in water as we look for the light on the horizon. I go to sleep, thinking of the stars above, and wondering why there is a working pump in a dried-out oasis.

The next day I drive to Al Ain, for an appointment with Mohsen Sherif, the Chairman of the Civil and Environmental Engineering Department, a Professor of water resources and the Director of the Water Resources Masters Programme in the UAE University. He explains that the United Arab Emirates uses 5.2 billion cubic metres of water a year. It has an allowance of five billion. Roughly, this is made of up of four billion of groundwater, which is used for agriculture, of one billion from six major desalination plants, which is for drinking, and the rest from recycled waste water and surface water, trapped in dams.

Twenty years ago it used less than half of this. Two decades back, when the Gulf was rich with oil, but hadn't yet imported western habits of consumption, the region used about 12 billion cubic litres a year. Now it goes through 32 billion. The massive increase is because these hot nations wanted to buy into the idea of global civilisation. The important elements of law, bureaucracy and urban life had been here for millennia, but the bar on what constituted success and modernity had been raised. So they built cities modelled on Los Angeles, and planted water-intensive crops like wheat and rice, and a forest in the desert, and fitted baths and washing machines and swimming pools. They looked at their new cities, compared them to the great metropolises of the world, and realised they needed green stuff – so they planted trees and lawns.

Mohsen says:

> This is good for the environment, for the beauty of the city, so you can see when you go to Dubai, there are many areas which are green, this develops the city. The green areas are part of the environmental situation, so of course to have a developed country you need to have some green area to a certain extent.
>
> A green city is a civilised one, a modern one, and so the money was spent and the water poured and the plants grew. The result is that Arabic habits of how to live in an arid environment were discarded.

The effect of this is

> ... most of the aquifers are depleting, and there is a notable decline in the ground water level in the last two decades, since the start of agriculture development.[61]

A date tree beside the new highway irrigated by groundwater is possibly a good use of the resource. It protects the road from sand drifts and erosion, while producing a crop. Much of the rest of the UAE's adventures in farming have been less productive. A lurch towards wheat merely sucked up water at great expense, while the vegetables that are grown are of such poor quality they end up as animal feed.

Mohsen talks about how these mistakes can be rectified:

> you need to focus on the crops that have been there for the centuries, the ones really suitable to the region. Also, we must end flood irrigation.

This is a lesson not just for the UAE, but for the Gulf.

> We have to make better agriculture choices, this is an important decision, that has to be taken not only in the UAE, but all the Gulf Co-operation Countries, because you cannot compete with other countries with plenty of water in the cultivation of rice or wheat. When Saudi Arabia started to cultivate wheat on a large scale, this came at the expense of water resources in the region and this

> affected the water resources in the UAE, because of the continuity of the groundwater resources in the region, so depleting the aquifer in one part will affect other countries.

Mohsen's vision is of a world co-operating region by region. So the Gulf, or Arab nations, will sort out who can best grow what, and share the spoils:

> You can import of course. I believe UAE should focus on the Arab world, because these kind of crops can be grown in Sudan, or Iraq, where water is available, so that they can count on neighbouring countries.

He looks to current co-operation between the Arab world, and the expanding European Union, as proof that regional management of water usage is beginning to happen. If a region can't cope, then he suggests importing food from elsewhere.

> There is no need to grow high-water crops, if you can bring it from India and Pakistan, from other countries with high water availability. I support the cultivation of crops according to water availability.

Mohsen finds optimism in the very fact that Dubai's water has rocketed – 'Our habits have changed just in the last 20 years, and we can change back in 20 years', he laughs, but is serious when he suggests that consumption could fall from around 650 litres per day to 250 litres per day. His main tool would be to end government subsidies on water and give it a market value, while stopping the wasteful use in agriculture. He also thinks that desalination can provide future needs if production costs drop, observing that the price has already fallen from around $3.5 dollars a litre in 2000, to $1 in 2008. Gulf states are debating the introduction of nuclear-powered desalination plants.

Our conversation ended. Mohsen Sherif and I shook hands, and smiled goodbyes. I walked out of the building, and into the alien heat. The car was thick with cooked air. As I drove off, I thought how Mohsen Sherif's plan to feed his native people depended on wheat and rice in India and Pakistan, and of the dried-up Indus, and the exhausted ground water of the subcontinent.

The road was a flat strip of tarmac through desert, and then, like a whip that has been cracked, it danced up through strange pockmarked mountains. I thought of how everyone's water plans relied on everyone else, but how so many were in the same trouble. Desalination may help some, as will a cut in consumption and better recycling of dirty water. However, Dubai's boulevards, its green proof of being developed, may have to wither and brown. The idea that being first world means using lots of water is, like the housing developments in the sea, built on sand.

At the coast I arrive at Fujirah, where circular oil depots are clustered on the harbour wall. Oil is pumped out to the tankers, while cargo ships unload luxury goods for the home market. Oil is the reason for the UAE's transformation in recent decades, but it isn't the only commodity in this region.

Here, in this ugly port, is the answer to the odd pumping at Liwa. The oasis in the desert is running dry, so the people get desalinated water from 250 km away in Fujirah. A pipeline takes it inland, to Al Ain, and then to Liwa. Because the desalination plants must operate at full capacity to meet peak demand, there are times when there is more water than is needed. Mindful of the Emirates' current water deficit, officials have found a new place for this excess supply. The pumps at Liwa are working to put the dollar-a-litre liquid back into the aquifers. As they grow rich pumping oil up, the money goes to refill nature's ancient water reservoirs – just in case the region isn't so co-operative after all.

'Water is directly related to life. Oil is not. So you can treat it differently. You cannot count on any other party for the delivery of water', Mohsen had said as I was leaving, 'It will decide the future of the nation, because you cannot sustain any development without water.'

The sun beats down. It is hot and dry. The problems of the UAE may seem distant, but they are the problems of the world. I think there is a water war, and it has already started. It is to this generation what the First World War and Second World War represented to their generations. It is the war for the future of civilisation and human survival on this planet. I'm not trying to make some clever point equating German nationalism with climate change and water scarcity, but I do believe, in terms of the enormity of the challenge and the threat it poses, we have our Third World War. Unpalatable as it is to a modern attitude that rejects blamewe will have to take sides, in the debate and we will have to take action. National boundaries and flag loyalty will come to mean nothing.

Unlike every other war, no nation or empire will win. Victory will be the continuation of civilisation, which in this context means the constant struggle to maintain and improve the conditions of living. Defeat is the unknown.

End Notes

1 PHILIP BALL'S BOOK H_2O is strongly recommended for explaining the science of water in an engaging style. He makes the point that the behaviour of water is exceptional in many respects; its oddness is its value.

2 John Reader's book *Man on Earth* offers this assessment:

> It is tempting to think of civilisation as an end, rather than a means, of human existence. But that would be wrong. There is no primary law that drives people towards civilisation as a mode of living. Civilisation is a very impressive demonstration of human ingenuity applied to the problems of fulfilling human requirements, but, in the story of human evolution, it is actually a very recent innovation. If the upright stance of human beings is taken to be a first indication of humanity, then the trail of fossil footprints uncovered at Laetoli shows that people have existed for at least 3.6 million years. If the manufacture of stone tools is regarded as a first sign of man's inventive potential then the earliest known stone tools (from Olduvai Gorge) show that people were already on the road to civilisation 1.9 million years ago. Civilisation as a way of life made its first appearance less than 5,000 years ago, however, which means that for 99.86 per cent of the time that human-like beings have been walking upright, and for 99.74 per cent of the time that has passed since humans first demonstrated a talent for technical innovation, people managed perfectly well without civilisation.

3 For some, civilisation means very bad things. As Ronald Wright puts it, 'As cultures grow more elaborate, and technologies more powerful, they themselves become ponderous specialisations – vulnerable, and in extreme cases, deadly.'

4 The movement of water has long fascinated thinkers. Pliny the Elder had suggested the ocean penetrated crevices in rocks and was carried through underground passages, filtering into fresh water, before emerging in higher land. After Leonardo da Vinci, Pierre Perault's 'Traite de

l'origine des fontaines', published in Paris in 1721, first described the hydrological cycle, based on a hypothesis set out in antiquity, that rivers are evaporated seawater condensed into rain, which collects in porous earth and wells up from impervious rock bed.

5 A Deluge, *1517* drawing in the Royal Collection, UK.

6 ... A late Palaeolithic child snatched from a campfire and raised among us now would have an even chance of earning a degree in astrophysics or computer science. To use a computer analogy, we are running 21st century software on hardware last upgraded 50,000 years ago or more, Steven Mithen the expert on pre-civilised man.

7 The archaeologist is Ofer Bar-Yosef, as quoted by Steven Mithen.

8 As Vernon Scarborough says, 'I suspect it (civilisation) evolves with many variables – it's really hard to identify a trigger. Partly just dating things is fuzzy – it can come down to where you put your spade.'

9 Interview with the author, January 2009.

10 That is one version of this story. Another tells it the other way round. Man learnt to control the fresh water from the Euphrates and the Tigris, as a result food became easier to find, so more people gathered to the area, and a city emerged. Or, Man learnt to channel water but realised it was a time consuming, tough job which required lots of people. It was in a bid to accommodate the labour needed to keep the irrigation channels open that a vast settlement was born. Interview with the author, January 2009, referring to the work of Guillermo Algaze of the University of Chicago.

11 This is important. A language based on many symbols is exclusive – it takes a great deal of time to learn and understand. One of relatively few symbols, capable of being arranged to describe many things, can be understood by many more. Writing becomes a democratic skill at this point, even if few master it.

12 Interview with the author February 2009.

13 As quoted by Fred Pearce in his book *The Dammed*.

14 The author Charles C. Mann says conscripts built dams, terraces, and irrigation canals, that everything was owned by the state. There was no money, and no markets, as the grain was distributed by the

state. They produced a surplus and Europeans found stuffed warehouses, and the belief is they eradicated hunger.

15 As highlighted in an interview with Vernon Scarborough.

16 *Landscape and Memory* by Simon Schama is very strong on this, and he uses the phrase 'magic of the Nile'.

17 As stated by Fred Pearce, p13, *The Dammed.*

18 Quoted by Sandra Postel in her book *Pillar of Sand.*

19 I'm not claiming this as an exact turning point, and retain a healthy doubt that such a milestone exists at all, but I have a sense that a shift occurs. Tony Wilkinson humoured me on this point and suggested 1,000 BC as the time when mankind gets bolder in his water schemes.

20 A. Trevor Hodge gives these details.

21 This is the opinion of A. Trevor Hodge

22 This myth is convoluted. Osiris was hated by his brother Set, his Egyptian name, known as Typhon by Diodorus. Typhon has a chest made to Osiris's dimensions and invites his brother to get in, saying whoever fits the bejewelled box can have it. Osiris is no sooner lying down than Typhon seals the chest with molten lead and throws it into the Nile. The coffin is washed ashore at Byblos, where Isis finds it. It has grown into a tamarisk tree, which has been cut down and turned into a supporting column for the king of Phoenicia's house. After various further magical twists and turns, Isis is able to return Osiris to Egypt. However, Typhon is enraged and hacks Osiris's body into 16 parts. Typhon scatters Osiris's body around, throwing his penis into the water. Isis is able to reassemble most of her brother's body, but not the phallus, which pumps its seed into the water, making it fertile. It is not that Osiris impregnates the water, but the water impregnates the land. The effect of his penis ending up in the Nile is that the river becomes a torrent of semen.

23 Plutarch has another shot at the myth in 'De Iside et Osiride', from the fifth book of his *Moralia*. This suggests Osiris ends up in 14 pieces. It adds an explanation why the river pike and sea bream are forbidden foods, because both are said to have eaten some of the discarded penis. To Plutarch, the Nile is the 'effusion of Osiris', confirming this sense

that the waters were literally seminal. Typhon is 'dry, fiery and arid', a barren thing, a god of the lifeless. When Osiris is in his coffin, the water vanishes and Egypt is thrown into anxiety. When Isis reconstitutes the dismembered Osiris, the river is rich again.

24 First century BC – mosaic on the floor of Aula Absidiata of the temple Fortuna Primigenia in the town of Palestrina 15 km south of Rome, as referred to in Schama.

25 Sextus Julius Frontinus, 'De Aquaeductu' 1 16, as translated by A. Trevor Hodge.

26 While Mesopotamia and Egypt are cultures which arise out of water control, the idea that water was at the beginning occurs in other myths. The first people in Japan to leave records, the Ainu, tell a story of how there was a muddy sludge at the beginning, which a heavenly bird divided into earth and water. In Mongolia, the Altaic tribes believed two black geese flew over water and one, a sort of male devil character, fell in and was forced to pull up the earth from the depths by the other, who represents God. This myth from the centre of the Asian land mass is echoed in one from North America, where the Apache Indians thought that mankind started beneath the water, in a kind of sub-aquatic nether world, only to emerge above the surface. Across the planet, people devise religious tales that begin with murky water and out of which emerges everything else.

However these myths are not the ones that take hold. They are tales which explain the existence of the world and man, but do not justify the works of civilisation. The religions that do offer a justification for man's dominance are, for obvious reasons, the ones associated with cultures which were powerful and expansive.

27 Simon Schama notes that Christianity starts out against the Nile, which represents sin and God's punishment. The Jordan is the river of Christianity – a clear, rushing line – carrying the chosen directly downstream. The Jordan gave water to the parched, unlike the Nile which sat around, feeding the sinners. This stems from the Essene cult near Qumran, the source of early Christian belief.

28 This translation from Ronald Wright's *A Short History of Progress*.

29 As quoted in *Gardens of Persia*.

30 *Gardens of Persia* offers this description of the qanat and their importance.

> On the (Iranian) plateau almost the only source of water is the snow which falls on the high peaks in winter and, since the seventh century BC, has been harnessed to make possible both the cultivation of fields and gardens... The ancient Persians and Medes, immigrants from the Russian steppes in around 1,000 BC... invented an elaborate system of underground conduits to take melting snow to the orchards and fields in the dry land of the plain. Since water in open irrigation canals on the high Iranian plateau evaporates very quickly in the summer heat, the ancient Persians built a network of underground aqueducts. This was done by sinking a shaft to the permanent subterranean water level at the base of the hills; from there a tunnel was dug to carry water to where it was needed. At intervals of 15m/50ft or so, further shafts were dug for removing spoil and to provide air for the underground workers. It required skill to achieve a straight line and for the precise 'tilt' for gravity to propel the flow of water. The excavator's guide was sometimes the shadow cast by a candle. Channels were lined with stone or tile in areas of particular porous soil and the tunnels, dug with the simplest of tools, could run for as many as 40km/25 miles.

31 As quoted p36, *A Little History of British Gardening*.

32 This is taken from the plans for St Gall, as redrawn by the Reverend Willis.

33 This may seem a cheat, as the main task was to hold back seawater, and this book is about fresh water. My justification for the importance of this is that it was necessary to get rid of the salt water if the fresh water was to stand a chance to be of agricultural use. By reclaiming land from the sea, they were claiming land for crops, to be nourished by clean water. As such, the distinction between sea water and fresh water, while obvious to people at the time, mattered less than the united interest to control all water.

34 Simon Schama's *Embarrassment of Riches* is excellent on the Dutch Empire, and the source of much material for this chapter.

35 With its 16 putti, representative of the sixteen cubits which was the mark of an ideal flood by the Egyptian river.

36 As quoted in *Conquest of Nature* by David Blackbourne, which is a very good reference work on this topic.

37 Tristram Hunt's book *Building Jerusalem* provides most of the quotations in this chapter.

38 The first century poet Antipater of Thessalonica on the mythical and practical value of water wheels, quoted in Lewis Mumford's *The Myth of the Machine, vol 1; Technics and Human Development.*

39 Interview with the author, January 2009.

40 Interview with the author, January 2009.

41 As quoted in *Water, a natural history.*

42 As quoted by Sandra Postel in *Pillar of Sand.*

43 All the following quotations from the ceremony are drawn from *Empires of the Word.*

44 The historian Jared Diamond offers eight categories for ecocide, as he calls it – why societies kill themselves off. These are deforestation and habitat destruction, soil problems, such as erosion or salinisation, water management problems, over-hunting, over-fishing, effects of introduced species, human population growth and the increased impact of people. I strongly recommend reading his book *Collapse* for further detail on these.

45 The eight categories are far from exclusive – they overlap. This underlines how hard it is to get a single definitive cause to ecocide. The reasons behind the collapse of a civilisation add a further layer of complexity. Diamond gathers his reasons for environmental degradation and combines them with other social factors, arriving at five reasons why civilisations collapse. These are environmental damage, climate change, hostile neighbours, friendly trade partners and responses to environmental problems.

46 Waldemar Kaempffert writing in the 1950 edition of *Popular Mechanics*, as reproduced in *The Times*, 1 September 2003.

47 The dimensions were used in an article in *The Guardian* newspaper 31 July 2007.

48 Adam Hart-Davis in *The Independent* magazine, 2004.

49 This quotation was used by *The Guardian* newspaper, 15 December 2007.

50 Interview with the author.

51 1976 UNESCO report, as quoted by A. Trevor Hodge.

52 These figures are given by Fred Pearce in his book *When the Rivers Run Dry.*

53 These figures are from Sandra Postel, *Pillar of Sand.*

54 22March 2006 John Vidal, *The Guardian.*

55 *The Guardian*, 8 May 2008.

56 *The Guardian*, 10 November 2006.

57 *The Skeptical Enviornmentalist* p. 56

58 It has long been a fantasy of civilisation that a source of power would be found which was inexhaustible and perpetual. In the 14th century, Villand de Honnecourt concocted an idea for a perpetual motion machine. Leonardo da Vinci had a shot. The Jesuit priest Johannes Taisnerius attempted a machine that would move without stop using magnets. The alchemist Cornelius Drebbel is alleged to have built the model. The patent office of England records an application for a perpetual motion machine in 1635. Three hundred years later, there were over 600 patents registered. None have worked. The last hundred or so can't claim to be more sophisticated.

59 *The Party's Over*, quoting 'The Coming Hydrogen Economy', *Fortune* magazine.

60 Quoted in *The Party's Over*, speaking at the Hyforum, Munich September 2000.

61 All quotes from Mohsen are from an interview with the author in July 2008.

Index

Luath Press Limited

committed to publishing well written books worth reading

LUATH PRESS takes its name from Robert Burns, whose little collie Luath (*Gael.*, swift or nimble) tripped up Jean Armour at a wedding and gave him the chance to speak to the woman who was to be his wife and the abiding love of his life. Burns called one of 'The Twa Dogs' Luath after Cuchullin's hunting dog in Ossian's *Fingal*. Luath Press was established in 1981 in the heart of Burns country, and is now based a few steps up the road from Burns' first lodgings on Edinburgh's Royal Mile.

Luath offers you distinctive writing with a hint of unexpected pleasures.

Most bookshops in the UK, the US, Canada, Australia, New Zealand and parts of Europe either carry our books in stock or can order them for you. To order direct from us, please send a £sterling cheque, postal order, international money order or your credit card details (number, address of cardholder and expiry date) to us at the address below. Please add post and packing as follows: UK – £1.00 per delivery address; overseas surface mail – £2.50 per delivery address; overseas airmail – £3.50 for the first book to each delivery address, plus £1.00 for each additional book by airmail to the same address. If your order is a gift, we will happily enclose your card or message at no extra charge.

ILLUSTRATION: IAN KELLAS

Luath Press Limited
543/2 Castlehill
The Royal Mile
Edinburgh EH1 2ND
Scotland
Telephone: 0131 225 4326 (24 hours)
Fax: 0131 225 4324
email: sales@luath.co.uk
Website: www.luath.co.uk